SpringerWienNewYork

AF608980

CISM COURSES AND LECTURES

Series Editors:

The Rectors
Giulio Maier - Milan
Jean Salençon - Palaiseau
Wilhelm Schneider - Wien

The Secretary General
Bernhard Schrefler - Padua

Executive Editor
Carlo Tasso - Udine

The series presents lecture notes, monographs, edited works and proceedings in the field of Mechanics, Engineering, Computer Science and Applied Mathematics.
Purpose of the series is to make known in the international scientific and technical community results obtained in some of the activities organized by CISM, the International Centre for Mechanical Sciences.

INTERNATIONAL CENTRE FOR MECHANICAL SCIENCES

COURSES AND LECTURES - No. 455

LIGHT GAUGE METAL STRUCTURES RECENT ADVANCES

EDITED BY

JACQUES RONDAL
UNIVERSITY OF LIEGE

DAN DUBINA
TECHNICAL UNIVERSITY OF TIMISOARA

SpringerWien NewYork

This volume contains 168 illustrations

This work is subject to copyright.
All rights are reserved,
whether the whole or part of the material is concerned
specifically those of translation, reprinting, re-use of illustrations,
broadcasting, reproduction by photocopying machine
or similar means, and storage in data banks.
© 2005 by CISM, Udine
Printed in Italy
SPIN 11405207

In order to make this volume available as economically and as rapidly as possible the authors' typescripts have been reproduced in their original forms. This method unfortunately has its typographical limitations but it is hoped that they in no way distract the reader.

ISBN 978-3-211-25258-1 SpringerWienNewYork

PREFACE

In recent years, it has been recognized that both cold-formed steel and aluminium alloy sections can be used effectively as primary framing components. In what concerns cold-formed steel sections, after their primarily application as purlins or side rails, the second major one in construction is in the building envelope. Options for steel cladding panels range from inexpensive profiled sheeting for industrial applications, through architectural flat panels used to achieve a prestigious look of the building. Light steel systems are widely used to support curtain wall panels. Cold-formed steel in the form of profiled decking has gained widespread acceptance over the past fifteen years as a basic component, along with concrete, in composite slabs. These are now prevalent in the multi-storey steel framed building market. Cold-formed steel members are efficient in terms of both their stiffness and strength. In addition, because the steel may be even less than 1 mm thick, the members are light weight. The already impressive load carrying capabilities of cold formed steel members will be enhanced by current work to develop composite systems, both for wall and floor structures.

Recent studies have shown that because the coating loss for galvanised steel members is sufficiently slow, and indeed slows down to effectively zero, a design life in excess of 60 years can be guaranteed. The production of economic coated steel coils has also given interesting solutions to architectural demands increasing the range of use of cold-formed sections. Higher yield stress steels are also becoming more common for the fabrication of cold-formed sections.

However, the use of high strength steels and thinner sections leads inevitably to complex design problems, particularly in the field of structural stability and joints. In recent years, stainless steel profiles and aluminium alloy profiles have also been used increasingly as structural members.

The aims of the "Advanced Professional Training on Light Gauge Metal Structures – Recent Advances" organized at the International Centre for Mechanical Sciences in Udine, June 3-7, 2002, were to review recent research and technical advances, including the progress in design codes, related to the engineering applications of light gauge metal sections made in carbon, high strength and stainless steel, as well as aluminium alloys.

The lectures include also a review of the new technologies for connections of light gauge metal members. Main advanced applications, for residential, non residential and industrial buildings and pallet rack systems are also covered.

This monograph is a revised version of the lecture notes. However, the lectures given by F.M. Mazzolani on the aluminium structural design have not been included in this monograph because a full CISM monograph, edited by F.M. Mazzolani (CISM Courses and Lectures n° 443, 2003) has been entirely dedicated to the use of Aluminium-Alloys in structures. The other lectures have been prepared by :

- *J.M. Davies, The University of Manchester, England;*
- *D. Dubina, Technical University of Timisoara, Romania;*
- *R. Laboube, University of Missouri-Rolla, USA;*
- *K.J.R. Rasmussen, University of Sydney, Australia;*
- *J. Rondal, University of Liege, Belgium.*

The editors wish to thank warmly these colleagues for the excellence of the work performed during the preparation of this advanced professional training, which is well reflected in this monograph.

Special thanks are also due to the CISM Rector, Prof. M.G. Velarde, the CISM Secretary General, Prof. B. Schrefler, the Executive Editor of the Series, Prof. C. Tasso, and to all the CISM staff in Udine.

Jacques Rondal
Dan Dubina

CONTENTS

Chapter 1: Introduction to Light Gauge Metal Structures

J. Rondal

Department of Mechanics of Materials and Structures, University of Liege, Belgium
E-mail: Jacques.Rondal@ulg.ac.be

1.1 Historical considerations

The use of cold-formed steel members in building construction began in the mid of the eighteenth century in United States and United Kingdom. However such steel members were not widely used as structural members until around 1946 and the publication of the first edition of the "Specification for the Design of Light Gage Steel Structural Members" by the American Iron and Steel Institute (AISI). Since that period, thousands of researches in the field have led to a wide use of cold-formed metal elements in all types of buildings.

If, in the past, cold-formed products were mainly used as secondary components in steel or concrete structures, there is now a wide marked for cold-formed structural elements.

These structural elements are used as single members like columns, beams or purlins but also as components of industrialized building systems. In these systems, the cold-formed elements play frequently a multifunctional role leading to economy and simplicity of the structure. Sometimes, they make the traditional steel skeleton unnecessary or, at least, they contribute largely to its load bearing capacity.

For example, the combination of cold-formed members and sheeting can be such that instability phenomena are prevented, leading to a space covering function and an improvement of the resistance.

1.2 Peculiarities of cold-formed steel members

In general, cold-formed steel structural members provide the following advantages in building construction (Yu, 1985):

- as compared with thicker hot-rolled shapes, cold-formed light members can be manufactured for relatively light loads and/or short spans;
- unusual sectional configurations can be produced economically by cold-forming operations and, consequently, favourable strength-to-weight ratios can be obtained;
- nestable sections can be produced, allowing for compact packaging and shipping;
- load-carrying panels and decks can provide useful surfaces for floor, roof, and wall construction, and in other cases, they can also provide enclosed cells for electrical and HVAC conduits;
- load-carrying panels and decks not only withstand loads normal to their surfaces, but they can also act as shear diaphragms to resist force in their own planes if they are adequately interconnected to each other and to supporting members.

Compared with other materials such as timber and concrete, the following qualities can be realized for cold-formed steel structural members:

- lightness;
- high strength and stiffness;
- ease of prefabrication and mass production;
- fast and easy erection and installation;
- substantial elimination of delays due to weather;
- more accurate detailing;
- nonshrinking and noncreeping at ambient temperature;
- uniform quality;
- economy in transportation and handling.

The combination of the above-mentioned advantages can result in important cost saving during construction.

However, because cold-formed members are usually thin-walled, special care must be given to the design. Compared to classical hot-rolled sections, they are characterized by some peculiarities, e.g.:

- large width to thickness ratios;
- singly symmetrical or unsymmetrical shapes;
- unstiffened or partially unstiffened parts of sections;

which can lead to difficult buckling problems :

- combined torsional and flexural buckling;
- local plate buckling;
- distorsional buckling;
- interaction between local and global buckling, …

Also connections must be designed with care because the thickness of the members can lead to local failures.

For these reasons, dedicated specifications have been published in United States firstly, and after in Europe, Australia and in other countries to cover these important questions.

1.3 Recent Advances

In recent years, it has been recognized that both cold-formed steel and aluminium alloy sections can be used effectively as primary framing components. In what concerns cold-formed steel sections, after their primarily application as purlins or side rails, the second major one in construction is in the building envelope. Options for steel cladding panels range from inexpensive profiled sheeting for industrial applications, through architectural flat panels used to achieve a prestigious look of the building. Light steel systems are widely used to support curtain wall panels. Cold-formed steel in the form of profiled decking has gained widespread acceptance over the past fifteen years as a basic component, along with concrete, in composite slabs. These are now prevalent in the multi-storey steel framed building market. Cold-formed steel members are efficient in terms of both their stiffness and strength. In addition, because the steel may be even less than 1 mm thick, the members are light weight. The already impressive load carrying capabilities of cold formed steel members will be enhanced by current work to develop composite systems, both for wall and floor structures.

Recent studies have shown that because the coating loss for galvanised steel members is sufficiently slow, and indeed slows down to effectively zero, a design life in excess of 60 years can be guaranteed. The production of economic coated steel coils has also given interesting solutions to architectural demands increasing the range of use of cold-formed sections.

As younger products, cold-formed steel sections are more open to development than classical hot-rolled profiles (Davies, 2000).

An important trend is the use of higher quality steels with an increased yield stress. The steel used actually for mass-produced products such as purlins, sheeting and decking has a yield stress in the range 280 to 600 N/mm^2. This trend is likely to continue in the future. However, the applications of high strength steels are limited by stiffness considerations in many situations.

The use of high strength steel leads inevitably to a reduction of the thickness of the profiles and to complex local stability problems. To improve the load stability of the sections, complex shapes have been developed with more folds and stiffeners.

Important progresses have also been made in the rolling and forming technology (Peköz, 1999). Modern rolling lines are computer controlled from the design office so that not only highly accurate complex shapes of precise lengths be produced to order but also holes, perforations, web opening for services can be punched in precise positions during the rolling process.

In recent years, stainless steel profiles and aluminium alloy profiles have also been used increasingly as structural members.

This Advanced Professional Training aims to review recent research and technical advances, including the progress in design codes, related to the engineering applications of light gauge metal sections made in carbon, high strength and stainless steel, as well as aluminium alloys.

References

Davies, J.M. (2000). Recent Research Advances in Cold-Formed Steel Structures. Journal of Constructional Steel Research, 55, 267-288.

Peköz, T. (1999). Possible Future Developments in the Design and Application of Cold-Formed Steel. ICSAS 99, 4th International Conference on Light-Weight Steel and Aluminium Structures. Espoo, Finland, 20-23 June 1999.

Yu, W.W. (1985). Cold-Formed Steel Design, J. Wiley and Sons, New-York.

Chapter 2: Peculiar Problems in Cold-formed Steel Design Part 1

D. Dubina

Department of Steel Structures and Structural Mechanics, Civil Engineering and Architecture Faculty, "Politehnica" University of Timisoara, Romania
E-mail: dubina@constructii.west.ro

2.1 Elements

2.2.1. Cold-formed steel sections: linear profiles, cladding and sheeting panels

In recent years, cold-formed steel sections started to be used effectively as primary framing components. In what concerns cold-formed steel sections, after their primarily applications as purlins or side rails, the second major one in construction is in the building envelope. Options for steel cladding panels range from inexpensive profiled sheeting for industrial applications, through architectural flat panels used to achieve a prestigious look of the building. Light steel systems are widely used to support curtain wall panels. Cold-formed steel in the form of profiled decking has gained widespread acceptance over the past fifteen years as a basic component, along with concrete, in composite slabs. These are now prevalent in the multi-storey steel framed building market. Cold-formed steel members are efficient in terms of both their stiffness and strength. In addition, because the steel may be even less than 1 mm thick, the members are light weight. The already impressive load carrying capabilities of cold-formed steel members will be enhanced by current work to develop composite systems, both for wall and floor structures.

The continuously increasing of cold-formed steel structures throughout the world is sustained by production of more economic steel coils particularly in coated form with zinc or aluminium / zinc coatings. These coils are subsequently formed into thin-walled sections by the cold-forming process. They are commonly called "Light gauge sections" since their thickness has been normally less than 3 mm. However, more recent developments have allowed sections up to 25 mm to be cold-formed, and open sections up to approximately 8mm thick are becoming common in building construction. The steel used for these sections may have a yield stress ranging from 250 MPa to 550 MPa (Hancock, 1997). The higher yield stress steels are also becoming more common as steel manufacturers produce high strength steel more efficiently.

Improving technology of manufacture and corrosion protection applied to cold-formed structural steelwork, provide competitiveness of resulting products and extend the area of new applications. Recent studies have shown that the coating loss for galvanized steel members is sufficiently slow, and indeed slows down to effectively zero, than a design life in excess of 60 years can be guaranteed (Owens, 2000).

Thin walled sections and high strength steels leads to design problems for structural engineers which may not normally be encountered in routine structural steel design. Structural instability of the sections is more likely to occur as a result of the reduced buckling loads (and stresses), and the use of higher strength steel which may make the buckling stress and yield stress of the thin-walled sections approximately equal (Hancock, 1997). Further, the shapes which can be cold-formed are often considerably more complex than hot-rolled steel shapes such as I-sections and unlipped channel sections. The cold-formed sections commonly have mono-symmetric or point symmetric shapes, and normally have stiffening lips on flanges and intermediate stiffeners in wide flanges and webs. Both simple and complex shapes can be formed for structural and non-structural applications as shown in Figure 1. Special design standards have been developed for these sections.

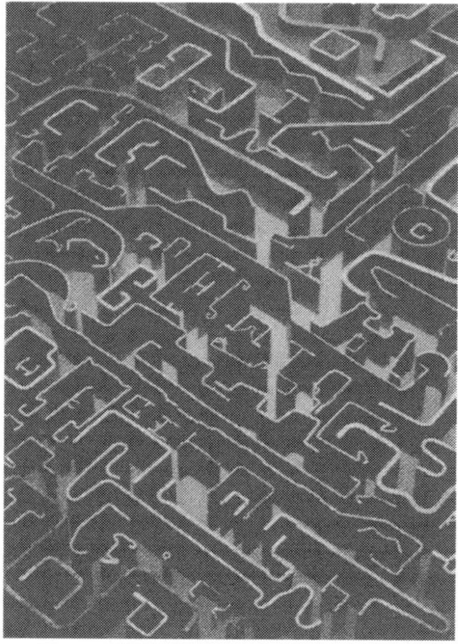

Figure 1. Collection of different cold-formed section shapes (Trebilcock, 1994).

Cold-formed members and profiles sheets are steel products made from coated or uncoated hot rolled or cold-rolled flat strip of coils. Within the permitted range of tolerances, they have constant or variable cross section.

Cold-formed structural steel members can be classified into two major types:

1. Long profile – individual structural framing
2. Cladding panels and sheeting decks

Individual structural members (bar members) obtained from so called "long products" include:

- single open sections, sown in Figure 2a;
- open built-up sections, Figure 2b;
- closed built-up sections, Figure 2c.

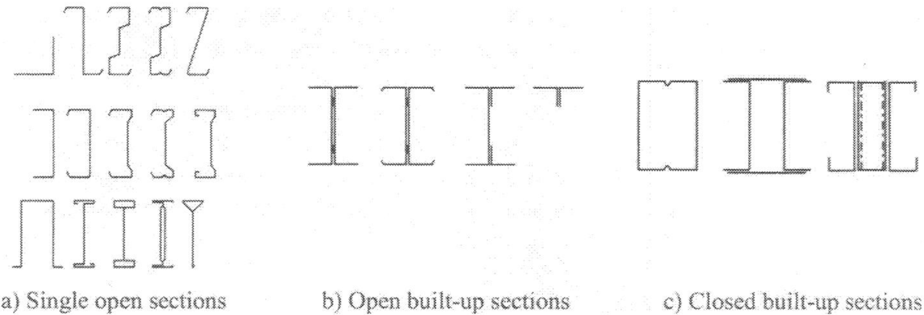

Figure 2. Typical forms of sections for cold-formed structural members.

Usual, the depth of cold-formed sections for bar members ranges from 50-70 mm to 350-400 mm, with thickness from 1 to 6 mm about.

Panel and decks are made from profiled sheets and linear trays (cassettes) shown in Figure 3. The depth of panel usually ranges from 20 to 200 mm, while thickness is from 0.4 to 1.2 (1.5) mm. They can be produced as flat or smooth curved shapes and can be used for roofing, wall cladding systems and load bearing deck panels.

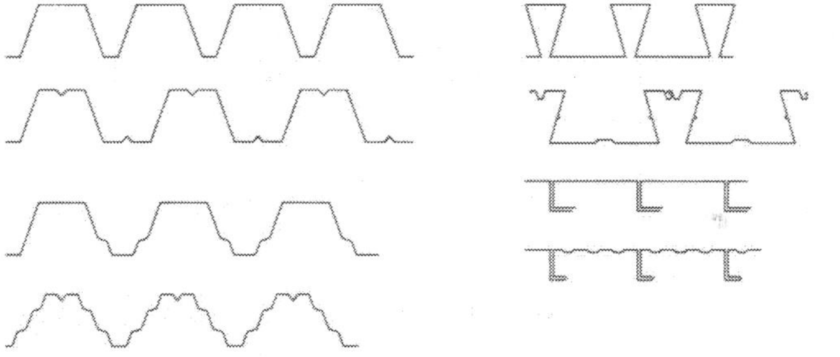

Figure 3. Profiled sheets and linear trays.

Smooth curved shape sheeting can be also produced by roll forming and bending special applications, like self-supporting arch and roof structures and also to provide a specific architectural appearance of facades.

In general, cold-formed steel sections provide the following advantages in building constructions (Yu, 2000):

1. As compared with thicker hot-rolled shapes, cold-formed light members can be manufactured for relatively light loads and/or short spans;
2. Unusual sectional configurations can be produced economically by cold-forming operations (Figure 1), and consequently favourable strength-to-weight ratios can be obtained;
3. Nestable sections can be produced, allowing for compact packaging and shipping;

4. Load carrying panel and decks can provide useful surface for floor, roof, and wall construction, and in other cases they can also provide enclosed cells for electrical and other conduits;
5. Load-carrying panels and decks not only withstand loads normal to their surfaces, but they can also act as shear diaphragms to resist force in their own panels if they are adequately interconnected to each other and to supporting members.

Compared with other materials such timber and concrete, the following qualities can be realized for cold-formed steel structural members:

1. Lightness;
2. High strength and stiffness;
3. Ability to provide long spans (up to 10 m, Rhodes, 1991);
4. Ease of prefabrication and mass production;
5. Fast and easy erection and installation;
6. Substantial elimination of delays due to weather;
7. More accurate detailing;
8. Non-shrinking and non-creeping at ambient temperatures;
9. Formwork unneeded;
10. Termite-proof and root-proof;
11. Uniform quality;
12. Economy in transportation and handling;
13. Non-combustibility;
14. Recyclable material.

Combination of above mentioned advantages can result in cost and erection time saving in construction.

2.1.2 Comparison with hot-rolled steel sections: Peculiar problems in Cold-formed steel design

The use of thin walled sections and cold-forming effects can result in special design problems, not normally encountered when tick hot-rolled sections are used. A brief summary of some special problems in cold-formed steel design is reviewed on the following.

Buckling strength of cold-formed members. Steel sections may be subjected to one of four generic types of buckling, namely local, global, distortional and shear (Davies, 2000). Local buckling is particularly prevalent in cold-formed sections and is characterised by the relatively short wavelength buckling of individual plate element. The term "global buckling" embraces Euler (flexural) and lateral-torsional buckling of columns and lateral buckling of beams. It is sometimes termed "rigid-body" buckling because any given cross-section moves as a rigid body without any distortion of the cross-section. Distortional buckling, as the term suggested, is buckling which takes place as a consequence of distortion of the cross section. In cold-formed sections, it is characterised by relative movement of fold-lines. The wavelength of distortional buckling is generally intermediate between that of local buckling and global buckling.

It is a consequence of the increasing complexity of section shapes that local buckling calculation is becoming more complicated and the torsional buckling takes on increasing importance.

Local and distortional buckling can be considered as "sectional" modes, and they can interact with each other as well as with global buckling (Dubina, 1996). Figure 4 shows single and interactive (coupled) buckling modes for a lipped channel section in compression. The results have been obtained using an elastic eigenbuckling FEM analysis. For given geometrical properties of member cross-section, the different buckling modes depend of buckling length. For shorter members, sectional buckling modes (L and D) are dominant, while for slender ones, the bar buckling modes (F and FT) prevail. Intermediate lengths are, generally, characterised by interactive sectional-bar buckling modes.

Sectional modes and their interaction with bar buckling ones do not appear in case of hot rolled sections.

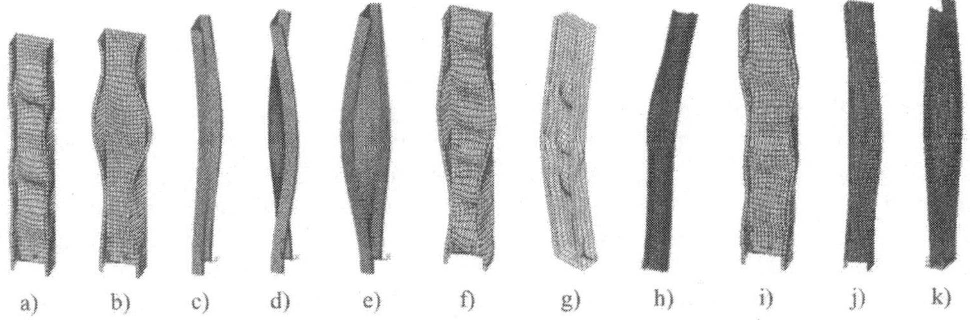

Figure 4. Buckling modes for a lipped channel in compression.
Single modes: (a) local (L); (b) distortional (D); (c) flexural (F);(d) torsional (T); (e) flexural-torsional (FT).*Coupled (interactive) modes*: (f) L + D; (g) F + L; (h) F + D; (i) FT + L; (j) FT + D; (k) F + FT

The effect of interaction between sectional and global buckling modes consists in increasing sensitivity to imperfections, leading to the erosion of theoretical buckling strength. In fact, due to the inherent presence of imperfection, buckling mode interaction always occurs in case of thin-walled members.

Figure 5 shows the difference in behaviour of a tick-walled slender bar in compression (Figure 5a), and a thin-walled one (Figure 5b). Both cases of ideal perfect bar and imperfect one are presented.

Looking to the behaviour of actual tick-walled bar it can be seen that it begins to depart from the elastic curve at point B when the first fibre reaches the yield stress and it reaches its maximum (ultimate) load capacity, N_u, at point C; after which it declines and the curve approaches the theoretical rigid-plastic curve asymptotically. The elastic theory is able to define the deflections and stresses up to the point of first yield and to define the load at which first yield occurs. The position of rigid-plastic curve determinates the absolute limit of load carrying capacity, for above it is a region in which the structures cannot carry a load and remain in a state of equilibrium.

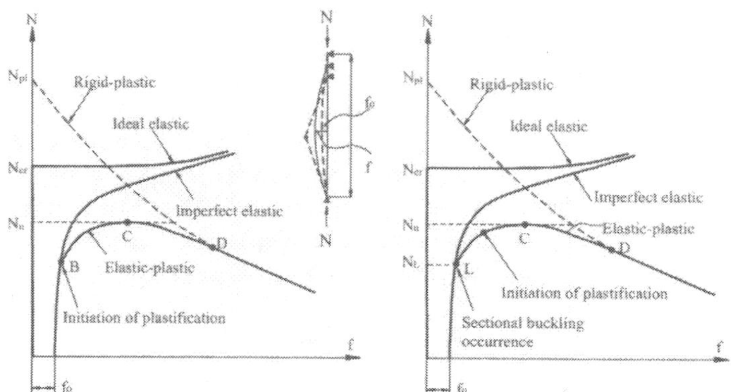

Figure 5. Behaviour of (a) slender tick-walled (hot-rolled section) and (b) thin-walled (cold-formed section) compression bar.

In case of thin-walled bar the sectional buckling, e.g. local or distortional buckling, occurs prior to the initiation of plastification. Sectional buckling is characterised by the stable post-critical path and bar does not fail when it occurs, but significantly lose from its stiffness. The yielding starts at the corners of cross-section, a few time before the failure of the bar, when sectional buckling changes into local plastic mechanism quasi-simultaneously with global buckling occurrence (Dubina, 2000).

In Figure 6 are shown the comparison between the buckling curves of a lipped channel member in compression, calculated according to ENV 1993-1-3, considering the full effective cross-section (e.g. no local buckling effect, which is generally the case of hot-rolled sections), and the reduced (effective) cross-section (e.g. when the local buckling occurs and interacts with global buckling).

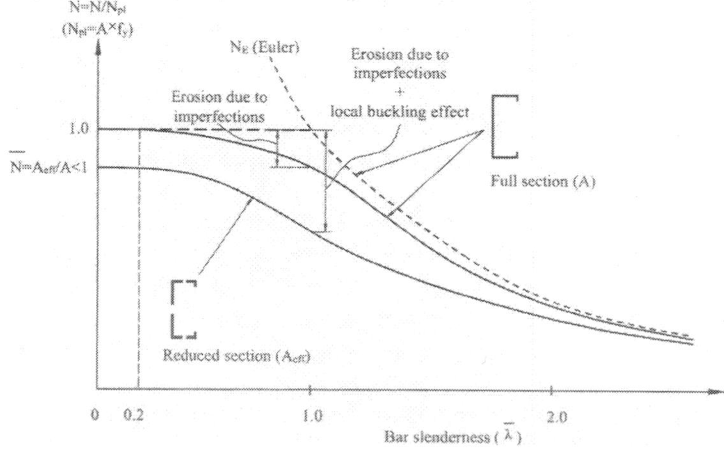

Figure 6. Effect of local buckling on the member capacity.

Web Crippling. Web crippling at points of concentrated load and supports can be a critical problem in cold-formed steel structural members and sheeting for several reasons. These are:

1. In cold-formed steel design, it is often not practical to provide load bearing and end bearing stiffeners. This is always the case in continuous sheeting and decking spanning several support points.
2. The depth-to-thickness ratios of the webs of cold-formed members are usually larger than hot-rolled structural members.
3. In many cases, the webs are inclined rather than vertical.
4. The intermediate element between the flange, on which the load is applied, and the web of a cold-formed member usually consists of a bend of finite radius. Hence the load is applied eccentrically from the web.

Web crippling is really a very peculiar feature of the behaviour of thin-walled cold-formed sections and special design provisions are included in design codes in order to manage this phenomenon, which doesn't occur in case of hot-rolled section.

Connections. Because of the wall thinness of cold-formed sections, conventional method for connection used in steel construction, such as bolting and arc-welding are, of course, available, but are generally less appropriate and emphasis is on special techniques, more suited to thin materials (Davies, 2000). Long-standing methods for connccting two elements thin material are blind rivets and self-drilling, self-tapping screws. Fired pins are often used to connect thin materials to a ticker-supporting member. More recently, press-joining or clinching technology (Predeschi, 1997), which is very productive, requires no additional components and causes no damage to the galvanizing or other metallic coating. This technology has been taken from the automotive industry, but actually it is successfully used in building construction. "Rosette" system is another innovative connecting technology (Makelainen and Kesti, 1999), proper to cold-formed steel structures.

Therefore, connection technology of cold-formed steel structures is representing one of their particular advantages, both in manufacturing and erection process.

Design assisted by testing. Cold-forming technology makes available production of unusual sectional configurations (see Figure 1). However, from the point of view of structural design, the analysis and design of such unusual members may be very complex. Structural systems formed by different cold-formed sections connected one to each other (like purlins and sheeting, for instance) can also lead to complex design situations, not entirely covered by design code procedures. Of course, numerical FEM analysis is always available, but even for some simply practical situations, modelling could be very complicate. For complex design problems, modern design codes permit to use testing procedures to evaluate structural performances. Testing can be used either to replace design by calculation or combined with calculation. Also one using numerical simulations, experimentally calibrated based numerical models are recommended. Only officially accredited laboratories, by competent authorities, are allowed to perform such tests and to delivery relevant certificates.

2.1.3 Ductility and plastic design

Usually, hot-rolled sections are of class 1 and 2, while cold-formed ones cold-formed sections are of class 4 or class 3, at the most. For this reason and also due to the effect of cold-forming by stress hardening, the cold-formed steel sections possess a low ductility and are not generally allowed for plastic design. Therefore, the previous discussion related to Figure 9b revealed the low inelastic capacity reserve for these sections, after the yielding was initiated. However, for members in bending, design codes allow to use the inelastic capacity reserve of their cross-section part which is working in tension. Moreover, because of their reduced ductility, cold-formed sections cannot dissipate energy in seismic resistant structures. Cold-formed sections can be used in seismic resistant structures because there are structural benefits coming from their reduced weight, but only elastic design is allowed and no reduction of shear seismic force is possible. Hence, in seismic design, a reduction factor q=1 has to be taken, as stated in EUROCODE 8 (ENV 1998, 1994).

2.1.4 Corrosion

The main factors governing the corrosion resistance of cold-formed steel sections is the type and thickness of the protective treatment applied to the steel and not the base metal thickness. Cold-formed steel has the advantage that the protective coatings can be applied to the strip during manufacture and before roll forming. Consequently, galvanized strip can be passed through the rolls and requires no further treatment.

Usually, steel profiles are hot dip galvanized with 275 gram of zinc per square meter (Zn 275), corresponding to a zinc thickness of 20 μm on each side. Hot dip galvanized is sufficient to protect the steel profiles against corrosion during the entire life of a building, if it was constructed in the correct manner. The most severe effects of corrosion on the steel occur during transport and storage outdoors. When making holes in hot dip galvanized steel framing members, normally no treatment is needed afterwards since the zinc layer a healing effect, i.e. transfers to unprotected surfaces.

Hot dip galvanizing is sufficient to protect the steel profiles against corrosion during the life of a building. The service life of hot dip galvanized steel studs was studied by British Steel and others (Burling P.M, 1990). The loss in zinc weight will be around 0.1g per m^2 per year indoors. A similar study was also carried out for steel floors above crawl spaces with plastic sheeting on the ground. Results showed that a zinc weight of 275g/ m^2 is sufficient to provide a durability of around 100 years.

Special attentions should be given to the cases in which different materials are intended to act compositely, if these materials are such that electrochemical phenomena might produce conditions leading to corrosion. Such phenomena are possible to appear when sheeting and fastener materials are different.

2.1.5 Fire resistance

Due to the small values of section factor (e.g. the ratio of the heated parameter to the cross-section area of the member) the fire resistance of unprotected cold-formed steel sections is reduced. In ENV 1993-1-2 (1995) simple calculation models are given by which the critical steel temperature and load bearing capacity easily can be found for different structures as beam and columns. However, these simple models are restricted to steel sections where the first order theory in a global plastic analysis may be used (ECCS TC3, 2001). Class 4 cross-sections do not fulfil the requirement and the subject has not been investigated much. The alternative is to use more general and more complicated calculation models and have to be verified by relevant test results. Very few attempts have been made to develop and verify calculation models for load bearing class 4 cross-sections in fire. In Klippstein (1979) and Gerlich (1995) calculation models are given for class 4 cross-sections, studs, in walls which have been verified by a number of fire tests. In Building Design: Fire protection (1993) the subject is described, principles and prediction methods are presented. In Ranby (1997), results show that a calculation according to ENV 1993-1-2 (1995) with reduced yield strength and plastic modulus is accurate enough also at elevated temperatures provided that the yield strength is taken as the 0.2 percent proof stress.

Table 1 shows comparatively the fire strength expressed in minutes for three different sections of which the area is approximately of comparable magnitude. One of these sections is built-up by back-to-back cold-formed lipped channels. The utilization factor is about 0.5. Eurocode 3-1.2 provisions and advanced FEM code SAFIR (Franssen, 2000) were used for calculation. The buckling was assumed to be prevented.

Sprayed cementations or gypsum based coatings can be employed for beams concealed behind a suspended ceiling. In load bearing applications, fire resistance periods of 30 minutes can usually be achieved by one layer of "special" fire resistant plasterboard, and 60 minutes by two layers of this plasterboard, which possesses low shrinkage and high integrity properties in fire. Planar protection to floors and walls provides adequate fire resistance to the enclosed sections, which retain a significant proportion of their strength, even at temperatures of 500°C.

In Light Gauge Steel Framing, the board covering of walls and floors can protect the steel against fire for up to 120 minutes, depending on the board material and the number of boards. The choice of insulation material, mineral wool or rock wool is also crucial to fire strength.

Table 1. Fire strength.

Member		Section	Am/V $[m^{-1}]$	FEM (SAFIR)	EC3-1.2	Difference
Unprotected	Columns	IPE220	254.7	11	11.5	-5%
		HEA160	234.7	11.5	10.8	6%
		][C300/3	480.4	9	9.3	-3%
	Beams	IPE220	254.7	11.9	11	8%
		HEA160	234.7	11.1	10.2	8%
		][C300/3	480.4	9.6	7.6	21%
Protected (Chartek 4 spray)	Columns	IPE220	254.7	34	29.8	12%
		HEA160	234.7	35.5	27.9	21%
		][C300/3	480.4	28	23.6	16%
	Beams	IPE220	254.7	35.9	28.5	21%
		HEA160	234.7	34.9	26.3	25%
		][C300/3	480.4	29.4	22.3	24%

Box protection to individual CFS sections used as beams and columns is provided in the same way as with hot rolled sections.

Non-load bearing members require less fire protection, as they only have to satisfy the "insulation" criterion in fire conditions. Ordinary plasterboard may be used in such cases.

However, using active fire protection and natural fire based design procedures seems to be the best solution in case of cold-formed steel sections.

References

Burling, P.M., et al (1990). Building with British Steel, *British Steel pls.*, England, No. 1.

Building Design Using Cold Formed Steel Sections: Fire Protection (1993), Publication P129, The Steel Construction Institute, Ascot, UK.

Davies, J.M. (2000). Recent Research Advances in Cold-Formed Steel Structures, General Report. *Journal of Constructional Steel Research* – Special Issue on Stability and Ductility of Steel Structures (Guest Editor D. Dubina), 55:1-3. 267-288.

Dubina, D. (1996). Coupled instabilities in bar members, General Report. In *Proceedings of the 2nd International Conference on Coupled Instabilities in Metal Structures-CIMS'96*, Imperial Colleague Press. 119-132.

Dubina, D. (2000). General Report – Session 3: Bar Members, in Coupled Instabilities. In *Proceedings of the 3rd International Conference on Coupled Instabilities in Metal Structures - CIMS'2000*, Imperial Colleague Press, 131-144.

ENV 1993-1-2 EUROCODE 3, Part 1.2 (1995). *Design of Steel Structures, Structural Fire Design*. European Committee for Standardization, Brussels.

ENV-1998-1 EUROCODE 8 (1999). *Design of structures for earthquake resistance Part 1. General Rules, Seismic Actions and Rules for Buildings*, CEN/TC250/SC8.

EN 1993-1-3 EUROCODE 3 (Draft of June 2001), *Design of Steel Structures, Part 1.3: General Rules, Supplementary Rules for Cold-Formed Thin-Gauge Members and Sheeting*, CEN/TC 250/SC3 – European Committee for Standardisation, Brussels.

ECCS TC3 (2001) European Convention for Constructional Steelwork - Technical Committee 3. *Model Code or Fire Design*, 1st Edition, Brussels: No. 111.

Franssen, J.M., Kodur, V.K.R., and Mason J. (2000). *User Manual for SAFIRT2001: A computer program for analysis of structures submitted to fire*. University of Liege.

Gerlich, J.T. (1995). *Design of Load Bearing Light Steel Frame Walls for Fire Resistance*, Fire Engineering Research Report 95/3, University of Canterbury, Christchurch, New Zeeland.

Hancock, G.J. (1997). Light Gauge Construction. *Progress in Structural Engineering and Materials*, Vol. I (I), 25-30.

Hancock, G.J. (1998). *Design of Cold-formed Steel Structures*. 3rd Edition, Australian Institute of Steel Construction, Sydney.

Klippstein, K.H. (1979). *Structural Performance of Cold Formed Steel Studs in Wall Exposed to ASTM E119 Fire Tests*, Final Report, US Steel Corp, Monroeville, Pennsylvania, US.

Law, S.C.W., and Hancock, G.J. (1987). Distortional Buckling Formulas for Channel Columns. *Journal of Structural Engineering, ASCE*, Vol. 113, No. 5, 1063-1078.

Makelainen, P., and Kesti, J. (1999). Advanced method for lightweight steel joining. *Journal of Constructional Steel Research*, No. 49, 107-116.

Owens, G. (2000). State-of-art report: Basic problems, design concepts and codification of steel and composite structures. Standards drafting: is there an opportunity for industry to improve value of its customers? General Report. *Journal of Constructional Steel Research*

– Special Issue on Stability and Ductility of Steel Structures (Guest Editor D. Dubina), 55:1-3. 7-27.

Predeschi, R.F., Sinha B.P., and Davies R. (1997). Advanced connection techniques for cold-formed steel structures. *Journal of Structural Engineering*, Vol. 123(2), 138-144.

Ranby, A. (1997). *Local Buckling of Thin Plates, Load Capacity in Case of Fire*, Report 188:3, Swedish Institute of Steel Construction, Stockholm, Sweden.

Rhodes, J. (1991). *Design of Cold-formed Steel Members*, Elsevier Applied Science, London and New York.

Trebilcock, P.J. (1994). *Building Design Using Cold-formed Steel Sections. An Architect's Guide.* SCI Publication P130. The Steel Construction Institute.

Yu, Wei-Wen (2000). *Cold-formed Steel Design.* 3rd Edition, John Willey & Sons, New York.

Chapter 2: Peculiar Problems in Cold-formed Steel Design Part 2

J. Rondal

Department of Mechanics of Materials and Structures, University of Liege, Belgium
E-mail: Jacques.Rondal@ulg.ac.be

2.2 Mechanical Properties and Imperfections

2.2.1 Introduction

Buckling and post-buckling of cold-formed members are rather difficult to predict due to material and geometrical nonlinearities. However, numerical techniques have reached a level of maturity such that many are now successfully undertaking ultimate strength analysis of cold-formed steel members (Schafer and Peköz, 1998).

The first condition to success of numerical simulations is not the theoretical formulation nor the solution technique but the knowledge of the initial state of a cold-formed steel member. Precise characterization of geometrical imperfections and residual stresses is largely unavailable. Also the distribution of the yield stress along the perimeter of the section is not uniform due to the cold-rolling process. A good knowledge of these fundamental quantities is necessary for reliable completion of advanced analysis or parametric studies of cold-formed steel members.

The aim of this chapter is to give information on the existing data related to these imperfections.

2.2.2 Mechanical properties of cold-formed steel members

Thin-walled steel sections are fabricated by means of cold-rolling of coils or press-braking of plates made in carbon steel. However, for these members which are now very frequently used in modern steel construction, the initial σ-ε relation of the steel is considerably changed by the cold-straining due to the manufacturing processes. Figure 7 shows the modification of the σ-ε diagram when a carbon steel specimen is first strained beyond the yield plateau and then unloaded.

If strain aging is now very rare, or at least limited, with modern steels, the cold-straining however modified the apparent σ-ε diagram which is pertinent for cold-formed steel members.

However, the strain-hardening can vary considerably along the cross-section due to the forming process. The apparent increase of the yield stress is more pronounced in the corners than in the flat faces as shown in Table 2.

Table 2. Increase of the yield stress

Forming process		Cold-rolling	Press-braking
Yield-strength (f_y)	Corners	High	High
	Flat faces	Moderate	—
Ultimate stress (f_u)	Corners	High	High
	Flat faces	Moderate	—

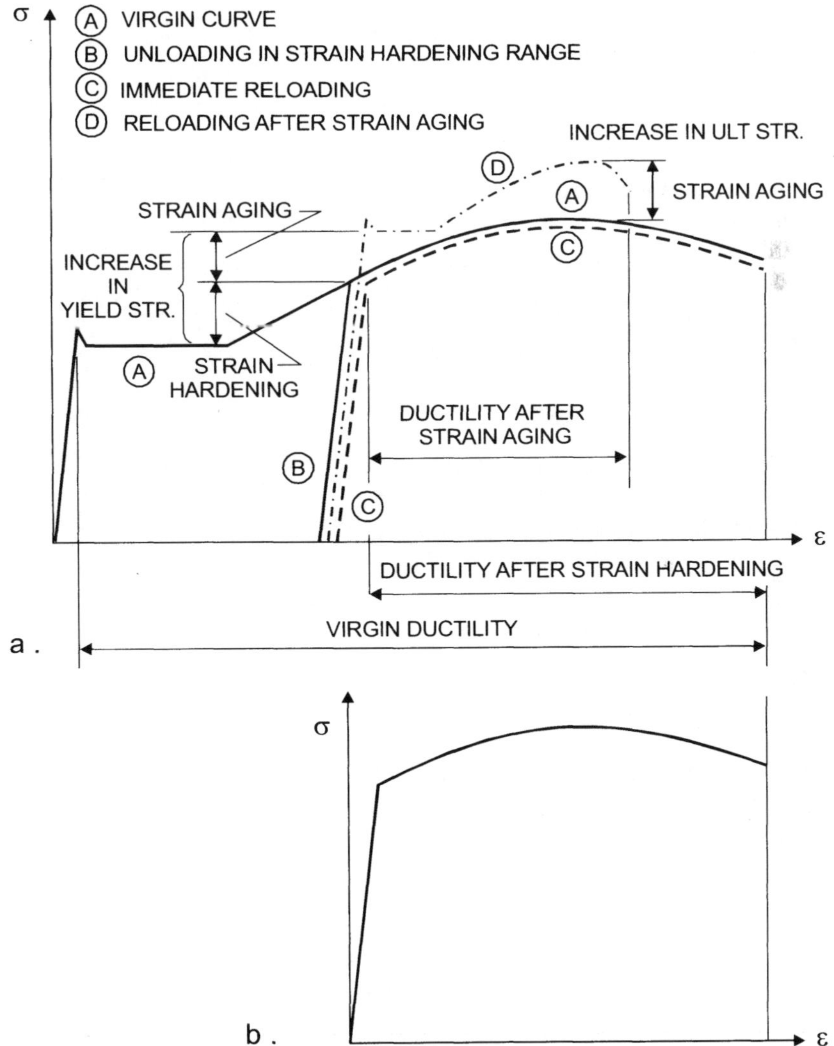

Figure 7. Effects of cold straining and strain aging on σ-ε characteristics of carbon steel (a – global σ-ε diagram; b – apparent σ-ε diagram for a cold-formed member)

Many authors have investigated the influence of cold work on the distribution of the yield strength along the cross-section.

Karren and Winter have proposed the following equation for the corner yield strength (Karren, 1967; Karren and Winter, 1967):

$$f_{yc} = \frac{kg}{(r/t)^h} \tag{1}$$

with

$$g = 0.945 - 1.315\,q \tag{2}$$

$$h = 0.803\,q \tag{3}$$

where t is the thickness of the sheet, r is the inside bend radius and k and q are the parameters of the hardening law which are given by :

$$k = 2.80 f_u - 1.55 f_{yb} \tag{4}$$

$$q = 0.225 \frac{f_u}{f_{yb}} - 0.120 \tag{5}$$

where f_u is the virgin ultimate strength and f_{yb} the virgin yield strength of the sheet.

With regard to the full-section properties, the average tensile yield strength may be approximated by using a weighted average as follows (Karren and Winter, 1967):

$$f_{ya} = A_c f_{yc} + (1 - A_c) f_{yb} \tag{6}$$

where A_c is the ratio of corner area to total cross-sectional area.

Eurocode 3, Part 1.3 gives the following formula to evaluate the average design strength f_{ya} of the full section. This formula is, in fact, a modification of formula (6) where a zone closed to the corner is considered as fully plastified:

$$f_{ya} = f_{yb} + (Cnt^2 / A_g).(f_u - f_{yb}) \tag{7}$$

where A_g is the gross cross sectional area and n is the number of 90° bends in the section, with an internal radius $r < 5t$ (Eurocode 3, 1996). In this relation, $C = 7$ for cold-rolling and $C = 5$ for other methods of forming.

The increase of the yield strength is however limited to a certain extent. Two formulae have been proposed for this limit:

$$f_{ya} \leq 0.5(f_{yb} + f_u) \tag{8}$$

or

$$f_{ya} \leq 1.25 f_{yb} \tag{9}$$

2.2.3 Geometrical imperfections

Geometrical imperfections refer to deviations from a so-called "perfect" geometry. Member imperfections included bowing, warping and twisting as well as local deviations characterized by dents and regular undulations in the plates.

Many researchers have measured geometrical imperfections of cold-formed steel structures.

Schafer and Peköz have analysed the frequency and amplitude of two types of local/sectional geometrical imperfections shown in Figure 8, where d_1 can be regarded as representative of local buckling, while d_2 characterises distorsional buckling (Schafer and Pekoz, 1998).

For simple analysis, these authors proposed to use the following values of the imperfections, for width to thickness ratio (w/t) less than 200 for type 1 imperfection and w/t less than 100 for the type 2, with a thickness less than 3 mm:

- type 1 :

$$d_1 \approx 0.006\, w \tag{10}$$

or

$$d_1 \approx 6t.e^{-2t} \tag{11}$$

The last formula is based on an exponential curve fit to the thickness of the profile.

- type 2 :

$$d_2 \approx t \tag{12}$$

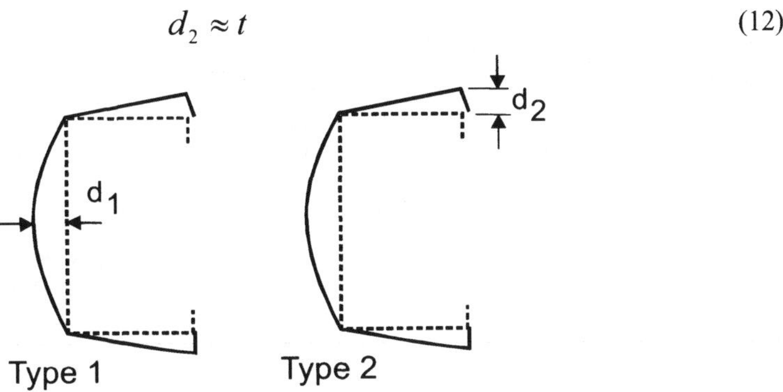

Figure 8. Definition of geometric imperfections

For advanced analysis, the best way to assess the influence of geometrical imperfections is to perform a direct probabilistic simulation considering both distribution and magnitude as random quantities. However this approach cannot be practically used due to the large amount of analyses required.

If little detailed information is known about imperfection variation along the length, maximum imperfections magnitudes are however available. Strength of a cold-formed steel member is particular sensitive to imperfections in the shape of its eigen modes. Knowledge of the amplitude of imperfections in the lowest eigen modes is often sufficient to characterize the influential imperfections. Maximum imperfections may be used to provide a conservative estimation of imperfection magnitude in a particular eigen mode.

Concerning the overall imperfection (member imperfection), the magnitude of L/1500 is generally considered as the mean of the statistical data for steel columns (Bjorhovde, 1972).

2.2.4 Residual stresses

Residual stresses are stresses that exist in materials independently of any external forces. They are produced in metals by most forming and fabrication processes of structural components: forging, rolling, welding, heat treating, grinding, machining, … These processes cause residual stresses by inducing plastic deformation of the metal through severe mechanical forces or temperature gradients. Localized permanent expansion or contraction of the metallic lattice in processes such as nitriding, carburizing or heat treatment which induce phase transformations can also generate residual stresses (Rondal, 1992):

Residual stresses can be classified in three categories:

1 In the first category macroscopic stresses are distributed over large areas and are due to plastic deformations or heat treatments.
2 In contrast, the residual stresses of the second category affect small areas such grains; they are due to the yield strength anisotropy in the crystals of materials undergoing plastic deformations.
3 Third category concerns residual microstresses; these are due to heat treatments and result from heterogeneous microstructures, like pearlite and ferrite, causing different elongation coefficients for various grains.

Amongst these three categories, the first one is the most important for structural purposes. Depending on their distribution and magnitude, residual stresses can have a significant negative effect on brittle fracture, fatigue, stress corrosion and buckling of steel members.

The distribution and the magnitude of the residual stresses depend very much on the fabrication process of the profile:

- in hot-rolled profiles, the residual stresses arise from non-uniform temperature distributions during the cooling stage of forming a hot-rolled shape. The part of the section which cools first is in residual compression, whilst the region which cools last is in residual tension. The residual stresses are mainly oriented in the direction of the rolling and have magnitudes rather constant through the thickness of the profiles (at least for profiles of moderate to medium thicknesses);
- in contrast to hot-rolled members, cold formed profiles are affected by deformational residual stresses in both longitudinal and transversal directions.

Major difficulties are met with when measuring residual stresses in cold-formed sections. On the one hand, measurements are allowed in a restricted number of points only at least for profiles of small dimensions. On the other hand, accurate measurements cannot be performed in the rounded parts of the sections, although these areas are subject to large residual stresses when cold-formed and are therefore worthwhile investigating. In the flat areas, measurements can be made more easily; but the accuracy is rather small because of the reduced amplitude of the residual stresses.

Any attempt to theoretical determining of residual stresses is therefore welcome. The theoretical methods aimed to this purpose will not depend, contrary to experimental methods, on the size of the specimens and will allow for parametric studies, with the result of a better understanding of the influence of the variables respectively and of the history of the onset of residual stresses.

Rondal has developed an analytical model able to calculate the distribution and amplitude of residual stresses in cold-formed steel sections (Rondal, 1987). However, the model shows that the variation of the stresses through thickness is rather complex and must be idealized for practical use.

Adequate computational modelling of residual stresses is not easy. Inclusion of residual stresses (at the integration points of the model for instance) may be complicated. Selecting an appropriate magnitude is made difficult by a lack of adequate data. As a result, residual stresses are often excluded altogether, or the stress-strain behaviour of the material is modified to approximate the effect of residual stresses.

In hot-rolled steel members residual stresses do not vary markedly through the thickness, in cold-formed members residual stresses are dominated by a “flexural”, or through thickness variation. This variation of residual stresses leads to early yielding on the faces of cold-formed steel plates. This important aspect of the load carrying behaviour is completely ignored unless residual stresses are explicitly considered in the analysis.

For cold-formed members, residual stresses are generally idealized as a summation of two types: flexural and membrane stresses.

Membrane residual stresses are generally rather low in cold-formed members but can be significative in the corners.

Compression membrane residual stresses cause a direct loss in compressive strength. Significant membrane residual stresses exist primarily in corner regions. Opposing this effect, the yield stress is elevated in corner regions due to the cold work of forming. If large membrane residual stresses are modelled in the corners or other heavily worked zones, then increased yield stress in these regions should be modelled as well. Conversely, if membrane residual stresses are ignored, then elevation of the yield stress should not be included. More study is needed to assess how much these two effects counteract really one another.

Large magnitude flexural residual stresses are regularly observed and flexural residual stresses larger than 50 % f_y are not uncommon. However these flexural residual stresses have a less detrimental influence on the buckling behaviour than membranar residual stresses (Costa Ferreira and Rondal, 1986).

Schafer and Peköz have proposed to use, in the numerical simulations, the average flexural residual stresses given in Figure 9.

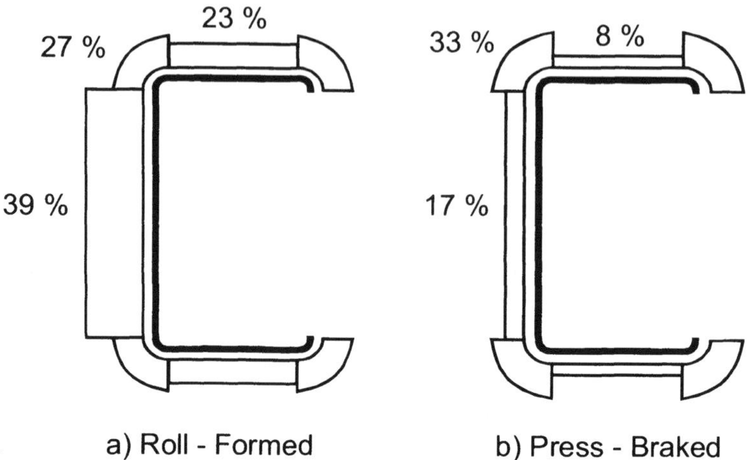

Figure 9. Average flexural residual stress as $\% f_y$ (compression inside, tension outside)

By comparison with the analytical model developed by Rondal and also experimental measures (Abdel-Rahman and Sivakumaran, 1996), for roll-formed profiles, the values proposed by Schafer and Peköz seems to be underestimated in the corners but largely overestimated in the web.

References

Bjorhovde, R. (1972). Deterministic and Probabilistic Approaches to the Strength of Steel Columns. Ph.D. Dissertation, Lehigh University, PA.

Costa Ferreira, C.M., Rondal, J. (1986). Influence of Flexural Residual Stresses on the Stability of Compressed Angles. *International Conference on Steel Structures, Recent Research Advances*. Budva, Yugoslavia, 28 September-1 October 1986, Vol. 1, 147-156.

Eurocode 3, Design of Steel Structures (1996). Part 1.3 – Supplementary Rules for Cold-Formed Thin Gauge Members and Sheeting. CEN, Brussel, Belgium.

Karren, K.W. (1967). Corner Properties of Cold-Formed Steel Shapes. *ASCE, Journal of the Structural Division*, Vol. 93, ST1.

Karren, K.W., Winter, G. (1967). Effects of Cold-forming on Light-Gage Steel Members. *ASCE, Journal of the Structural Division*, Vol. 93, ST1.

Rondal, J. (1987). Residual Stresses in Cold-Rolled Profiles. *Construction and Building Materials*, Vol. 1, 3, 150-164.

Rondal, J. (1992). Residual Stresses : Measurements Theoretical Predictions and their Use in the Design of Steel Structures. *Testing of Metals for Structures*, E. and F.N. Spon, London, 381-391.

Schafer, B.W., Peköz, T. (1998). Computational Modeling of Cold-formed Steel: Characterizing Geometric Imperfections and Residual Stresses. *Journal of Constructional Steel Research*, 47, 193-210.

Chapter 3: Recent Advances and Progress in Design Codes: Instability Problems

J. Rondal

Department of Mechanics of Materials and Structures, University of Liege, Belgium
E-mail: Jacques.Rondal@ulg.ac.be

3.1 Introduction

Cold-formed steel sections may be subject to three basic types of buckling:

- local buckling;
- distorsional buckling;
- global buckling.

The term global buckling (or long mode buckling) embraces flexural buckling, torsional and flexural-torsional buckling and lateral-torsional buckling. The half-wave length of these modes is of the order of magnitude of the length of the member. Global buckling is characterized by rigid body movements of the whole member such that individual cross-sections rotate and translate but do not distort in shape.

Local and distorsional buckling are sometimes called sectional buckling because they depend very much of the geometry of the cross-section. Local buckling is characterized by a relatively short half-wavelength of the order of magnitude of individual plate elements while the fold lines remain straight.

Distorsional buckling occurs at a half-wavelength intermediate to local and global mode buckling. The half-wavelength is typically several times larger than the largest characteristic dimension of the cross-section. Distorsional buckling involves both translation and rotation at the fold lines of a member leading to a distorsion of the cross-section.

Figure 1 illustrates these three modes of buckling for a member under pure compression.

Each of the three generic categories of buckling is capable of mutual interaction. Empirical models for the interaction of local and global buckling are included in most design codes but there is only few fundamental knowledge of the interaction of distorsional buckling with the other modes (Davies, 1999).

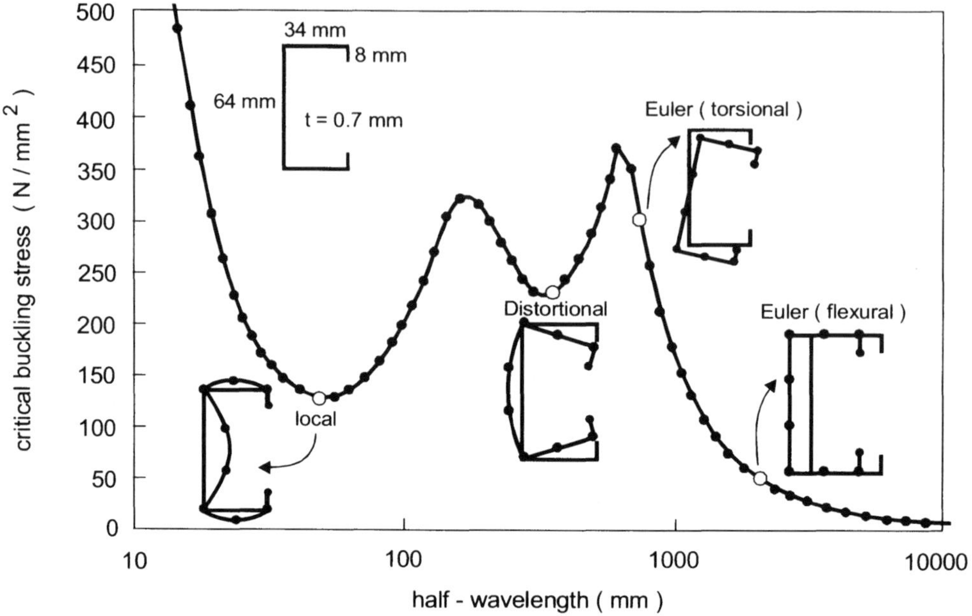

Figure 1. Buckling modes of a lipped channel

3.2 Local and Distorsional Buckling

3.2.1 Local buckling

Elastic local buckling is typically treated by ignoring any interaction between elements (flanges and web). Each element is considered independently and classical plate buckling equations based on isolated simply supported plates are generally used. The result of this approach is that each element of the section is predicted to buckle at a different stress. This approach, called the element model, can lead to rather conservative predictions. The critical stress of each element is given by the well-known Von Karman formula:

$$\sigma_{cr} = k_{\sigma} \frac{\pi^2 E}{12(1-\nu^2)} (\frac{t}{b})^2 \tag{1}$$

where b is the width of the considered element and k_{σ} is the buckling factor which depends of the type of element (double supported element called unstiffened element) and of the stress distribution on the element.

The buckling factor k_{σ} is equal to 4 for a stiffened element and to 0.43 for an unstiffened element under uniform compression and is given in the different specifications for non-uniform stress distributions.

In order to better predict the actual local buckling stress, semi-empirical methods that include element interaction have been proposed.

In the approach proposed by Schafer, expressions for k_σ are determined for both flange/lip and flange/web local buckling by empirical close-fit solutions to finite strip analysis results (Schafer, 2001):

- flange/lip : $$k_\sigma = -11.07\left(\frac{d}{b}\right)^2 + 3.95\left(\frac{d}{b}\right) + 4 \quad \text{if } d/b < 0.6 \tag{2}$$

- flange/web : $$k_\sigma = 4\left(\frac{b}{h}\right)^2\left[2 - \left(\frac{b}{h}\right)^{0.4}\right] \quad \text{if } h/b \geq 1 \tag{3}$$

$$k_\sigma = 4\left[2 - \left(\frac{b}{h}\right)^{0.2}\right] \quad \text{if } h/b < 1 \tag{4}$$

Batista has proposed an approach giving the entire member local buckling stress also based on close-fit solutions to finite strip analysis results (Batista, 1988):

- lipped channel :

$$k_\sigma = 6.8 - 5.8\left(\frac{b_2}{b_1}\right) + 9.2\left(\frac{b_2}{b_1}\right)^2 - 6.0\left(\frac{b_2}{b_1}\right)^3 \tag{5}$$

with $$0.1 \leq \frac{b_2}{b_1} \leq 1.0 \text{ and } 0.1 \leq \frac{b_3}{b_1} \leq 0.3$$

- plain channel :

$$k_\sigma = 4.02 + 3.44\left(\frac{b_2}{b_1}\right) + 21.76\left(\frac{b_2}{b_1}\right)^2 - 174.33\left(\frac{b_2}{b_1}\right)^3 + 319.94\left(\frac{b_2}{b_1}\right)^4 - 237.55\left(\frac{b_2}{b_1}\right)^5 + 63.60\left(\frac{b_2}{b_1}\right)^6 \tag{6}$$

with $$0 \leq \frac{b_2}{b_1} \leq 1.0$$

- hollow rectangular section :

$$k_\sigma = 6{,}56 - 5{,}77\left(\frac{b_2}{b_1}\right) + 8.56\left(\frac{b_2}{b_1}\right)^2 - 5{,}36\left(\frac{b_2}{b_1}\right)^3 \tag{7}$$

with $$0{,}1 \le \frac{b_2}{b_1} \le 1{,}0$$

Figure 2 shows clearly the influence of the interaction of the elements. For example, for a plain channel, k_σ increases first with the flange width up to $b_2/b_1 \approx 0.2$ and decreases after due to the prominent buckling of the flanges.

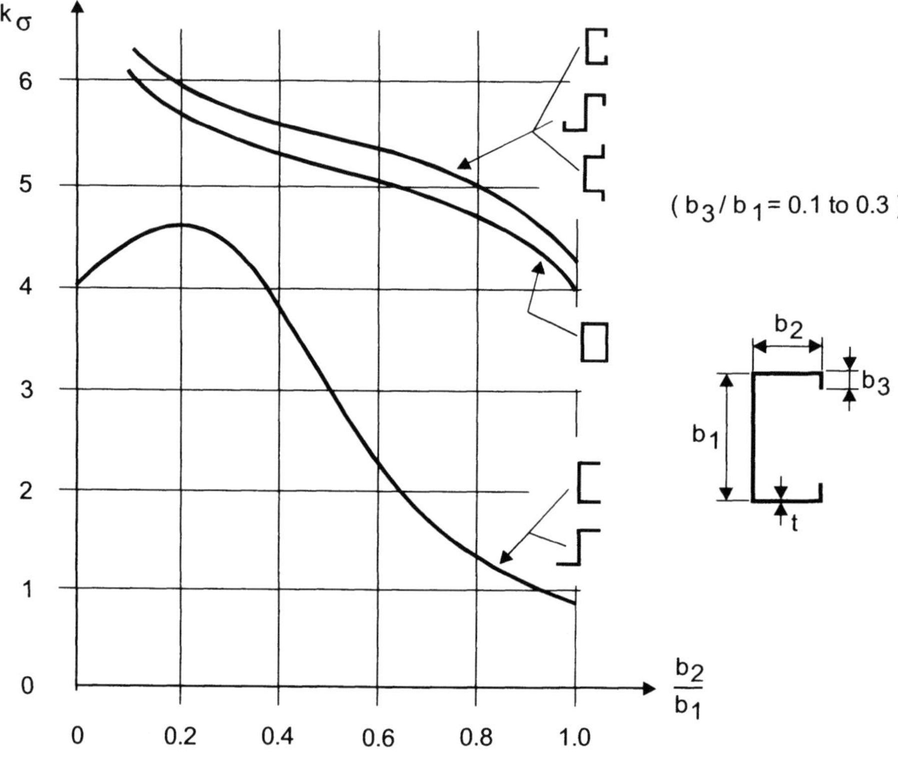

Figure 2. Buckling factor for some shapes of cold-formed profiles

Semi-empirical methods have however rather limited applications because results are only published for classical types of profiles. For more complex cross-sections, i.e. with more folds or internal stiffeners, it will be better to use numerical methods, such as the finite element method, the finite strip method or generalized beam theory (GBT) to determine the local buckling stress of an entire member. Computer programs, mainly based on the finite strip method and the generalized beam theory have been specially written for the determination of the local buckling stress of cold-formed steel members. However, it must be stressed that these computer programs gives also answers on the other modes of buckling such that distorsional buckling and global buckling.

A finite strip based program with very efficient pre- and post-processors written for cold-formed steel members can be downloaded freely on the web site of Schafer of the Johns Hopkins University (www.ce.jhu.edu/bschafer/cufsm). A commercial program for finite strip analysis is also distributed by the University of Sydney (www.civil.usyd.edu.au/case/THINWALL.html).

3.2.2 Distorsional buckling prediction

Distorsional buckling of compression members such as lipped channels usually involves rotation of each flange and lip about the flange-web junction in opposite directions as shown in Fig. 3. The web undergoes flexure at the same half-wavelength as the flange buckle, and the whole section may translate in a direction normal to the web also at the same half-wavelength as the flange and web buckling deformations. The web buckle involves single curvature transverse bending of the web. Distorsional buckling of compression members has been investigated in detail in Hancock (1985) mainly for sections used in steel storage racks, Lau & Hancock (1987) for a range of lipped channel sections, and by Kwon & Hancock (1992) for high-strength steel channel sections with intermediate stiffeners.

Distorsional buckling of flexural members such as plain channels and Z sections usually involves rotation of only the compression flange and lip about the flange-web junction as shown in Figure 3. The web undergoes flexure at the same half-wavelength as the flange buckle, and the compression flange may translate in a direction normal to the web, also at the same half-wavelength as the flange and web buckling deformations. The web buckle involves double curvature transverse bending of the web.

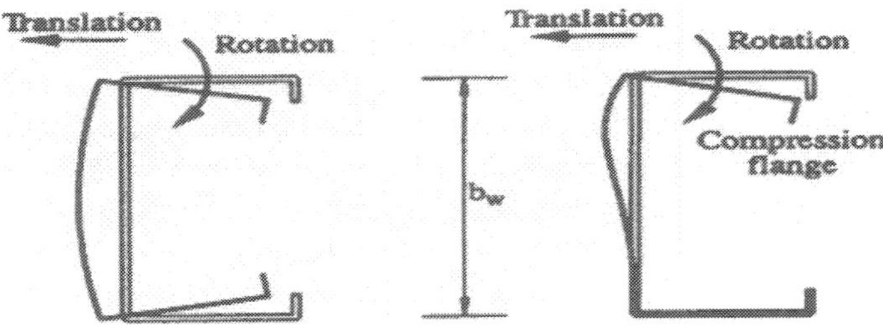

Figure 3.

A general model for the determination of the elastic distorsional buckling stress under axial compression has been originally developed by Lau and Hancock (1987). Figure 4 shows this analytical model which is based on a flange buckling where the flange is treated as a compression member restrained by a rotational and a translational spring. The rotational spring stiffness k_ϕ

represents the torsional restraint from the web and the translational spring stiffness k_x represents the torsional restraint to translational movement of the cross-section.

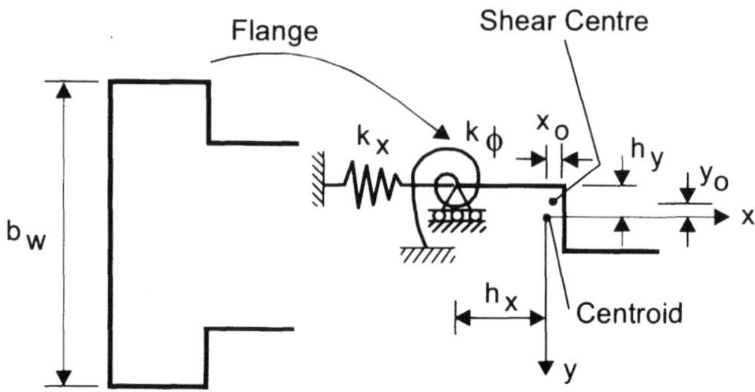

Figure 4. Analytical model for distorsional column buckling

Lau and Hanock (1987) showed that the translational spring stiffness k_x does not have much significance and the value of k_x was assumed to be zero. The key to evaluating this model is to consider the rotational spring stiffness k_ϕ and the half buckling wavelength λ, while taking account of symmetry. Lau and Hancock gave a detailed analysis in which the effect of the local buckling stress in the web and of shear and flange distortion were taken into account in determining expressions for k_ϕ and λ, the buckling length. This gives rise to a rather long and detailed series of explicit equations for the distorsional buckling stress. Not withstanding their cumbersome nature, these are now included in the Australian code.

A similar set of explicit equations has also proposed by Schafer (2001) and will be incorporated in future AISI Specifications.

The buckling behaviour of beams bent about the major axis differs from that of a compressed member in a number of respects.

Analytical expressions for the distorsional buckling of thin-walled beams of general section geometry under a constant bending moment about the major axis have been developed by Hancock (1995). These analytical expressions were based on the simple flange buckling model shown in Figure 5, together with an improvement proposed by Davies and Jiang (1996) in which the flange was again treated as a compression member with both rotational and translational spring restraints in the longitudinal direction. The rotational spring stiffness k_ϕ and the translational spring stiffness k_x represent the torsional restraint and translational restraint from the web respectively. In his analysis, Hancock again assumes the translational spring stiffness k_x to be zero.

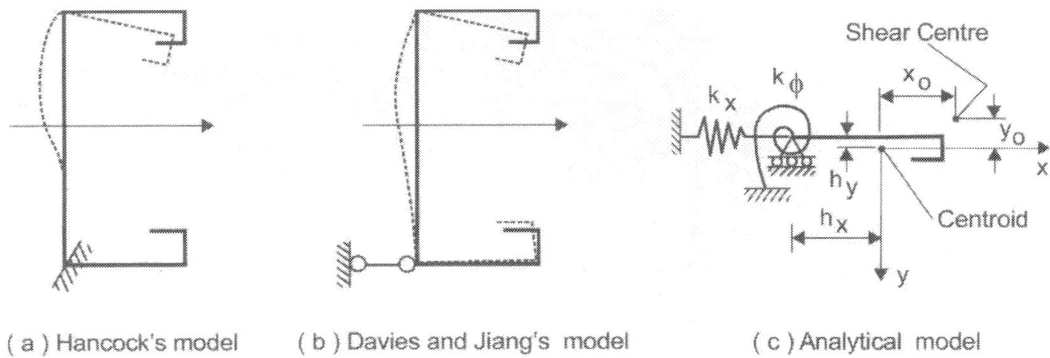

Figure 5. Analytical model for distorsional beam buckling

These beam models are, of course, directly analogous to the column model shown in Fig. 5. The only significant difference lies in the stiffness of the rotational spring and the necessary modifications to the design expressions for the rotational spring stiffness k_ϕ and the buckling length λ are given in Hancock's paper. This then leads to similar equations for the critical stress for distorsional buckling (Davies, 1999).

Explicit expressions for the prediction of the distorsional buckling of beams have also been proposed by Schafer and Peköz (1999).

3.2.3 Design strength under local and distorsional buckling

For the determination of the design strength under local buckling, effective widths and effective cross-section properties are generally used. The semi-empirical formula, due to Winter, is generally used in the specifications:

$$b_{eff} = \rho b \tag{8}$$

with:

$$\rho = 1.0 \text{ if } \overline{\lambda}_p \leq 0.673 \tag{9}$$

$$\rho = (1.0 - 0{,}22/\overline{\lambda}_p)/\overline{\lambda}_p \text{ if } \overline{\lambda}_p > 0.673 \tag{10}$$

in which the plate slenderness is given by :

$$\overline{\lambda}_p = \sqrt{\frac{f_y}{\sigma_{cr}}} = 1.052\frac{b}{t}\sqrt{\frac{f_y}{Ek_\sigma}} \tag{11}$$

The specifications give the value of k_σ for different stress distributions on the element and also the distribution of the effective parts of the section.

The reduced properties of effective plate elements in compression may then be combined with the full width of plate elements in tension to give an effective cross-section for use in strength calculations.

If the critical stress has been calculated for the entire member by means of semi-empirical formulae or with a computer program, the Winter's formula can be directly used at the level of the entire member by replacing in formula (11) the critical stress of the plate element by the critical stress of the entire member. The design strength under local buckling is then calculated by the relation:

$$f_L = \rho . f_y \tag{12}$$

For the determination of the design strength under distorsional buckling, Hancock (1985) has given expressions established on the base of effective widths for distorsional buckling.

Schafer (2001) proposes to calculate the design stress under distorsional buckling with the following relation:

$$f_D = f_y \left[1 - 0.25 \left(\frac{\sigma_{cr}}{f_y} \right)^{0.6} \right] \left(\frac{\sigma_{cr}}{f_y} \right)^{0.6} \tag{13}$$

$$\text{if} \quad \sqrt{\frac{f_y}{\sigma_{cr}}} > 0.561 \text{; else } f_D = f_y \tag{14}$$

where σ_{cr} is the elastic distorsional critical stress.

3.3 Global and Interactive Buckling

3.3.1 Buckling curves

In Eurocode3 (1993) an Ayrton-Perry formula is used for the calculation of the design strength of columns under compression (Rondal and Maquoi, 1979).

To take into account the interaction between local and global buckling, the calculation of the load bearing capacity has to be based upon the effective cross-section, calculated for uniform compression.

A member is subject to concentric compression if the line of action goes through the neutral axis of the effective cross-section under uniform compression. It the line of action does not go through the neutral axis of the effective cross-section, the member has to be checked for compression and bending.

The design buckling resistance $N_{b.Rd}$ with respect to flexural buckling shall be taken as:

$$N_{b.Rd} = \chi . A_{eff} . f_y / \gamma_{M1} \tag{15}$$

$$\chi = \frac{1}{\phi + \left[\phi^2 - \overline{\lambda}^2\right]^{1/2}} \text{ but } \chi \leq 1 \tag{16}$$

$$\phi = 0.5\left[1 + \alpha(\overline{\lambda} - 0.2) + \overline{\lambda}^2\right] \tag{17}$$

where:

A_g = area of gross cross-section

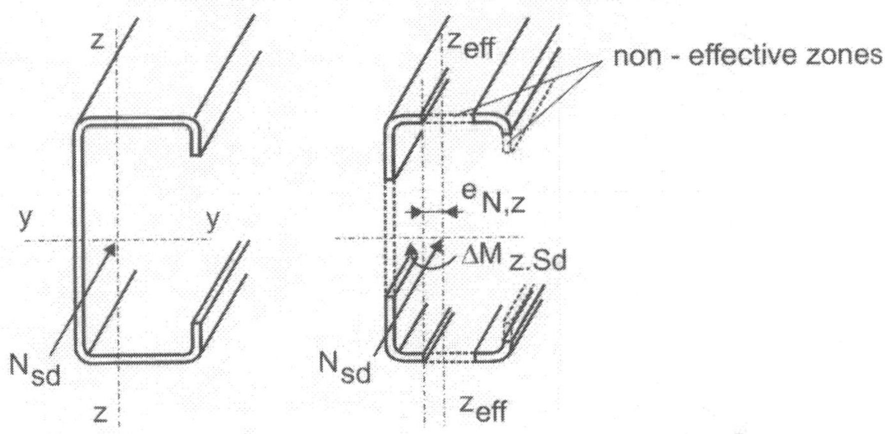

a) Gross cross-section b) Effective cross-section for uniform compression

Buckling about y-axis: $e_{N.y} = 0 \quad \Delta M_{y.Sd} = 0 \Rightarrow$ concentric compression

Buckling about z-axis: $e_{N.z} \neq 0 \quad \Delta M_{z.Sd} = N_{Sd} . e_{N.z} \Rightarrow$ bending and compression

Figure 6. Shift of neutral axis due to the effective cross-section

A_{eff} = area of effective cross-section (at uniform compression at yield stress level)

$\overline{\lambda} = \left[A_{eff} . f_y / N_{cr}\right]^{1/2} = (\lambda / \lambda_1) . \left[\beta_A\right]^{1/2}$

$\beta_A = A_{eff} / A_g$

N_{cr} = the elastic critical axial force for flexural buckling for the gross cross-section

λ = slenderness for the relevant buckling mode

$\lambda = l / i_g$ (either $\lambda_y = l_y / i_{g.y}$ or $\lambda_z = l_z / i_{g.z}$)

$\lambda_1 = \pi.\left[E / f_y\right]^{1/2}$

i_g = radius of gyration about the relevant axis (either $i_{g.y}$ or $i_{g.z}$) determined by using the properties of the gross cross-section

α = imperfection factor, depending on the appropriate buckling curve

The imperfection factor α corresponding to the appropriate buckling curve shall be obtained from Table 1.

Table 1. The α imperfection factor

Buckling curve	a	a	b	c	c
α	0.13	0.21	0.34	0.49	0.76

Buckling curve *b* is used for *C* and *Z* profiles and buckling curve *C* is used for U profiles and angles.

Similar relations are also given for lateral-torsional buckling of beams.

3.3.2 Erosion of the critical bifurcation load

On the basis of the erosion of critical bifurcation load theory, Dubina et al. have proposed, for simple and coupled instability modes, a different approach (Dubina, 1990).

This approach assumes that the two theoretical instability modes that couple in a thin-walled compression member, are the Euler overall instability mode, $\overline{N}_E = 1/\overline{\lambda}^2$ and the local instability mode described by means of Q factor, i.e. $\overline{N}_L = Q$ (Figure 7) where $Q = A_{eff} / A_g$.

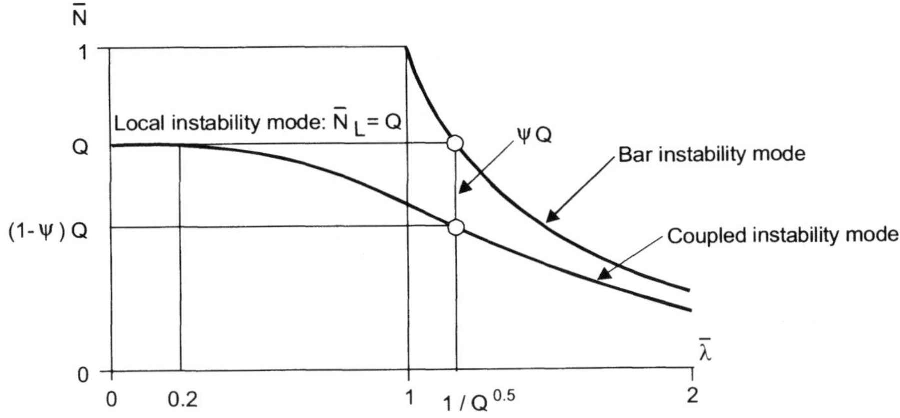

Figure 7. The Interactive Buckling Model based on the ECBL Theory

It must be underlined that $\overline{N}_L = Q$ is not the equation of the theoretical local buckling curve, but it can be assumed (in a simplified way, of course) as a level of the cross-section local buckling mode.

The maximum erosion of the critical load, due both to the imperfections and coupling effect is occurring in the coupling point, $\overline{\lambda} = 1/\sqrt{Q}$. The interactive buckling curve, $\overline{N}(\overline{\lambda}, Q, \psi)$ that pass through this point, is plotted in Figure 7; the corresponding value of ultimate buckling load is $\overline{N}(\overline{\lambda} = \frac{1}{\sqrt{Q}}, Q, \psi) = (1-\psi)Q$, where ψ is the erosion factor.

_It is now imposed that the Ayrton-Perry buckling curve must be equal to $(1-\psi)$ in point $\overline{\lambda} = 1$, because it corresponds to the maximum erosion of the member theoretical buckling curve (Figure 8):

$$\overline{N}(\overline{\lambda} = 1, \alpha) = \frac{1}{2}\left[2 + 0.8\alpha - \sqrt{(2 + 0.8\alpha)^2 - 4}\right] = 1 - \psi \tag{18}$$

that gives :

$$\alpha = \frac{\psi^2}{0.8(1-\psi)}, \tag{19}$$

respectively,

$$\psi = 0.4(\sqrt{5\alpha + \alpha^2} - \alpha) \tag{20}$$

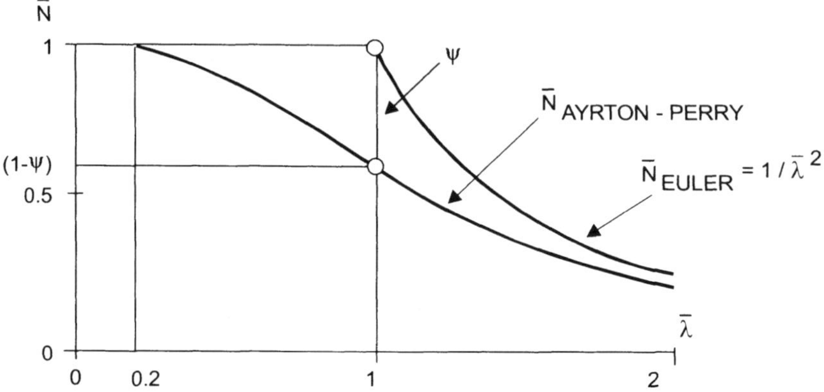

Figure 8. The erosion of bar buckling curve

When local buckling occurs prior the bar buckling, then:

$$\overline{N} = \frac{1+\alpha(\overline{\lambda}-0{,}2)+Q\overline{\lambda}^2}{2\overline{\lambda}^2} - \frac{1}{2\overline{\lambda}^2}\sqrt{\left[1+\alpha(\overline{\lambda}-0{,}2)+Q\overline{\lambda}^2\right]^2 - 4Q\overline{\lambda}^2} = (1-\psi)Q \quad (21)$$

and with $\overline{\lambda} = \dfrac{1}{\sqrt{Q}}$, it gives $\alpha = \dfrac{\psi^2}{1-\psi} \cdot \dfrac{\sqrt{Q}}{1-0.2\sqrt{Q}}$ (22)

This represents the new formula of α imperfection coefficient which should be introduced in the European buckling curves in order to adapt these curves to overall-local buckling.

Extensions of the ECBL approach to lateral-torsional buckling of thin-walled steel beams have also been proposed (Dubina, 1998).

3.3.3 Direct strength methods

Schafer and Peköz (1998) have recently proposed a new procedure which works only with the gross properties of a member and can take into account the interaction between local and global buckling but also the interaction between distorsional and global buckling.

Direct strength methods are the extension of column curves to other modes such as local and distorsional buckling. Formulae are proposed for beams (Schafer and Peköz, 1998) and for columns (Schafer, 2001). They are here only given for columns.

For local buckling, the following formula is proposed:

$$\frac{P_n}{P} = \left(1-0.15\left(\frac{P_{cr\ell}}{P}\right)^{0.4}\right)\left(\frac{P_{cr\ell}}{P}\right)^{0.4} \text{ for } \sqrt{\frac{P}{P_{cr\ell}}} > 0.776\text{, else } P_n = P \quad (23)$$

where:

P_n is the nominal capacity
P is the squash load ($P = P_y = A_g f_y$) except when interaction with other modes is considered, then $P = A_g f$ where f is the limiting stress of the interaction mode (global buckling for example)
$P_{cr\ell}$ is the critical elastic local buckling load ($A_g f_{cr\ell}$).

For distorsional buckling, the nominal capacity of the member is given by:

$$\frac{P_n}{P} = \left(1 - 0.25\left(\frac{P_{crd}}{P}\right)^{0.6}\right)\left(\frac{P_{crd}}{P}\right) \text{ for } \sqrt{\frac{P}{P_{crd}}} > 0.561 \text{, else } P_n = P. \qquad (24)$$

where:

P_n is the nominal capacity
P is the squash load ($P = P_y = A_g f_y$) when interaction with other modes is not considered, otherwise $P = A_g f$, where f is the limiting stress of a mode that may interact
P_{crd} is the critical elastic distorsional buckling load ($A_g f_{crd}$).

For global buckling, these authors proposed to use the following expressions:

$$P_n = A_g f_n \text{ for } \lambda_c \leq 1{,}5 f_n = \left(0.658^{\lambda_c^2}\right) f_y \text{ for } \lambda_c > 1{,}5 f_n = \left(\frac{0.877}{\lambda_c^2}\right) f_y \qquad (25)$$

where:

P_n is the nominal capacity

$\lambda_c = (f_y / f_e)^{1/2}$ and f_e is Euler buckling stress (min. of flexural and flexural-torsional, with appropriate braced lengths)
f_y is the yield stress.

References

Batista, E.M. (1988). Etude de la stabilité des profils à parois minces et section ouverte de type U et C. Ph.D. Thesis, University of Liege, Belgium.

Davies, J.M., Jiang, C. (1996). Design of Thin-Walled Beams for Distorsional Buckling. *13th Int. Specialty Conference on Cold-Formed Steel Structures*, St-Louis, Missouri, 1996, 141-153.

Davies, J.M. (1999). Modelling, Analysis and Design of Thin-Walled Steel Structures; ICSAS99, *4th International Conference on Light-Weight Steel and Aluminium Structures*, Espoo, Finland, 20-23 June 1999, 3-18.

Dubina, D. (1990). A new approach to the interaction of local and overall buckling in thin-walled cold-formed compressed members. *4th International Colloquium on Stability of Steel Structures,* Final Report, Budapest, Hungary, April 25-27, 1999, 412-419.

Dubina, D. (1998). Interactive Buckling Analysis of Thin-Walled Cold-Formed Steel Members. *Coupled Instabilities in Metal Structures*, Edited by J. Rondal, Springer, Wien, 1998.

Eurocode 3 (1993). Part 1.1. Design of Steel Structures; General Rules and Rules for Buildings.

Hancock, G.J. (1985). Distorsional Buckling of Steel Storage Rack Columns. *ASCE, Journal of Structural Engineering*, 111, 12, 2770-2783.

Kwon, Y.B., Hancock, G.J. (1992).Strength Tests of Cold-Formed Channel Sections undergoing Local and Distorsional Buckling, *ASCE, Journal of Structural Engineering , 117, 2, 1786-1803.*

Lau, S.C.W., Hancock, G.J. (1987). Distorsional Buckling Formulas for Channel Columns. *ASCE, Journal of Structural Engineering*, 113, 5, 1063-1078.

Rondal, J., Maquoi, R. (1979). Formulation d'Ayrton-Perry pour le flambement des barres métalliques. *Construction Métallique*, 4, 41-53.

Schafer, B.W., Peköz, T. (1998). Direct Strength Prediction of Cold-Formed Steel Members using Numerical Elastic Buckling Solutions. *14th International Specialty Conference on Cold-Formed Steel Structures,* St Louis, Missouri, 1998.

Schafer, B.W., Peköz, T. (1999). Local and Distorsional Buckling of Cold-Formed Steel Members with Edge Stiffeners Flanges. *ICSAS 99, 4th International Conference on Light-Weight Steel and Aluminium Structures*, Espoo, Finland, 20-23 June 1999, 89-97.

Schafer, B.W. (2001). Thin-Walled Column Design Considering Local, Distorsional and Euler Buckling. Structural Stability Research Council. Annual Technical Session and Meeting 2001.

Chapter 4: Recent Advances and Progress in Design Codes: Connections

Roger A. LaBoube

Department of Civil Engineering, University of Missouri-Rolla, USA
E-mail : laboube@umr.edu

4.1 Introduction

Because cold-formed steel members are fabricated from thin sheet steel, the potential limit states, or failure modes, differ from a similar connection in thicker hot-rolled steel members. Also, the thinner sheet steel offers the opportunity for the application of a broader array of connector. For example, in hot-rolled construction, only two types of connectors, a bolt or a weld, are routinely employed. However in cold-formed steel construction, bolts, welds and screws are routinely used, and pins and rivets are also often employed. Thus, connection design is more complex and challenging to the engineer. This paper and subsequent papers will explore the types and connectors and the key limit states that must be considered to ensure an adequate structural design.

4.2 General Design Rules

Design of cold-formed steel members and connections are governed by Eurocode 3: Design of Steel Structures, Part 1.3: General Rules, Supplementary Rules for Cold-Formed Thin Gauge Members and Sheeting (Eurocode 3-1.3, 1996).

Noteworthy is the limitation on the core thickness of the steel sheet. In Section 3.1.3, Eurocode 3 stipulates that when design is to be based on calculation given in Part 1.3, steel sheet must be within the following thickness ranges:

- for sheeting: $0.5 \text{ mm} \leq t_{cor} \leq 4.0 \text{ mm}$
- for members: $1.0 \text{ mm} \leq t_{cor} \leq 8.0 \text{ mm}$

where core thickness, t_{cor}, is the sheet thickness exclusive of zinc or organic coating.

Section 8.4 in the code (ENV 1993-1.8) summarizes the general rules that apply to the design of the four types of mechanical fasteners: bolts, screws, rivets, and pins. When determining the positions of fasteners, care should be given to place the fasteners as close as practical, but allowance must be given for satisfactory assembly and maintenance.

The behavior of connections having mechanical fasteners is extremely complex and highly indeterminate. Research has shown that the actual shear forces on fasteners in a group vary with the location of the fastener. However, because of the favorable ductility of steel connections, local stress concentrations do not detrimentally effect the structural performance of the member or its connection. To simplify design using mechanical fasteners, Section 8.4 states that the shear forces on individual mechanical fasteners is a connection may be assumed equal provided:

- the fasteners have sufficient rigidity, and

- shear is not the critical failure mode.

When design employs a bolt, the fastener resistance is usually defined by the grade of bolt. However, resistances for a screw, rivet, or pin are not stipulated by an industry grade. Thus, the shear or tension resistance for a screw, rivet, or pin must be determined by test in accordance with Section 9.

In fact, Eurocode 3, in an effort to not limit innovation or application of cold-formed steel members and connections permits the use of steel sheet thicker or thinner than the above limits provided that the load carrying capacity is determined by test in accordance with Section 9.

4.3 Connecting Devices

Section 3.2 of Eurocode 3 summarizes the types of connecting devices that are recognized by the standard. Section 3.2.1 states that bolts, nuts, and washers shall conform to the requirements of ENV 1993-1-1.

Section 3.2.2 states that screws, pins and rivets are may also be used for connecting cold-formed steel members. Section 3.2.2 also stipulates that screws may be of the thread-forming or thread-cutting type and may be either self-drilling or self-tapping.

Welding of cold-formed steel members shall conform to the requirements of ENV 1993-1-1, as stated in Section 3.2.3 of Eurocode 3, Part 1.3.

Because cold-formed members are typically formed from thin sheet, connecting devices are not necessarily limited to the conventional connectors. For example, in addition to the above named fastener types, thin sheets can be connected by press joints or clinches, self-piercing rivets or nails, and cold-formed seams.

Clinch joining is a process by which metal parts are connected by cold-forming. A punch is used to press the metal into a die with sufficient force to cause a portion of the metal to flow sideways thus forming a lock with the bottom sheet. The feature of the connection is that the metal itself provides the fastener (Light, 1999). The Rosette connector is a form of a clinch, but does not rely on the sideways material flow to accomplish the connection (Makelainen, 1998).

The self-piercing rivet or nail is made of high carbon steel and are heat treated to make it very hard, yet ductile. This enables the nail to penetrate steel but will not create a brittle failure. The nail has a ballistic shaped point and a deformed, knurled, surface. The nail is pneumatically driven into the steel member (Light, 1998).

A major short coming of a screw attached roof panel is the potential for leaks resulting from movement of the panel due to thermal forces. To provide a more weather tight membrane, roof panel manufacturers have developed the standing seam panel. The standing seam panels are interconnected at their side laps by seaming the panels together. The seaming is actually a cold-forming procedure that is accomplished on the construction site.

4.4 Bolted Connections

Although the same grade of bolt may be used for either cold-formed steel connections or hot-rolled, thicker sheet, connections, the behavior of the bolted connections may be different. The difference in behavior is attributed to the thinness of sheets used in cold-formed steel connections.

Eurocode 3, Part 1.3 lists six different strength grades of fasteners that may be used in cold-formed steel construction: 4.6, 5.6, 8.8, 4.8, 5.8, 6.8, and 10.9. The yield and tensile strengths for each bolt grade is given in Section 3.3.2.1 of Part 1.1 of the Eurocode 3.

In hot-rolled steel construction, bolted connections may be designed as either a bearing type connection or a slip resistant type connection. However, in cold-formed steel construction, only bearing type connections are used. The slip resistant type connection is not recognized in cold-formed steel construction primarily because of the difficulty to achieve the requisite pre-tension in the bolt. In fact, in the United States, pretensioning of bolts in not required. Research has demonstrated that the strength of a bearing type connection is independent of the level of bolt preload. Thus, installation must only ensure that the bolted assembly will not come apart during service. Experience has shown that bolts installed to a snug tight condition, that is no preload, do not loosen under normal building load conditions.

Holes for bolts may be drilled or punched, although punched is the preferred for speed of fabrication. The nominal clearance in standard holes shall be as follows:

- 1 mm for M12 and M14 bolts
- 2 mm for M16 to M24 bolts
- 3 mm for M27 and larger bolts.

When design is based on calculation, as summarized in Table 8.4 of Part 1.3, the thinner connected part or sheet must be equal to or greater than 1.25 mm.

Four general types of failure modes must be considered when designing a bolted connection. When insufficient end distance from the centre of the fastener to the adjacent end of the connected part in the direction of the load, e_1, a longitudinal shear failure may occur (Fig. 1a.). If sufficient end distance is provide, such that a longitudinal shear failure is prevented, bearing failure of the connected part may occur (Fig. 1b.). Although not a common limit state, fracture in the net section of the connected part must be evaluated (Fig. 1c.). Another infrequently occurring limit state, shear failure of the bolt, must also be investigated.

4.4.1 Longitudinal Shear

The minimum edge distance of each connected part, e_1, is determined by the following design equation:

$$F_{b,Rd} \leq f_u \, e_1 \, t / 1.2 / \gamma_{M2}$$

where f_u = ultimate tensile strength of the connected sheet, t = thickness of the thinner connected sheet, and $\gamma_{M2} = 1.25$.

4.4.2 Net Section Resistance

In Eurocode 3, Part 1.3, the design tension resistance on the net section of connected parts is based on the provisions of Sections C5.2 and 8.4 whichever is smaller.

Section C5.2 prescribes the limit state of yielding in the gross section as

$$N_{t,Rd} = f_{ya} \, A_g / \gamma_{MO}$$

where f_{ya} = average yield strength, A_g = gross area of the cross section, and γ_{MO} =1.10. In addition to the limit state on yielding in the gross section, Section C5.2 also limits $N_{t,Rd}$ to the fracture in the net section resistance as prescribed by Section 8.4.

Net section resistance is defined by the following equation:

$$F_{n,Rd} = (1+3r(d_o/u - 0.3))A_{net}\, f_u/\, \gamma_{M2} \leq A_{net}\, f_u / \gamma_{M2}$$

where r = number of bolts at the cross section / total number of bolts in the connections, and u = 2 e_2 with a limit of u ≤ p_2, the spacing centre-to-centre of fasteners in the direction perpendicular to the direction of load.

The equation for $F_{n,Rd}$ is based on tests of flat sheet connections. Thus, the quantity in the parentheses is a reduction factor to reflect that the stress is not uniform in the connection cross section. This non-uniform stress is called shear lag. Shear lag is more critical in cross sections having multiple elements, such as angles and C-sections. Eurocode 3 does not contain design rules for addressing shear lag in cross sections. For guidance the designer may refer to the American Iron and Steel Institutes specification (North, 2001).

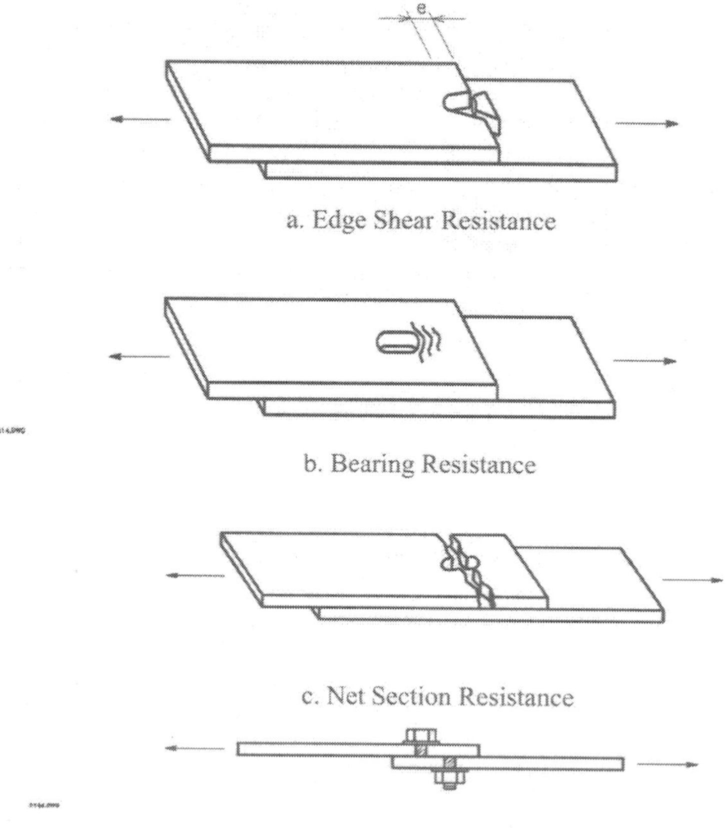

a. Edge Shear Resistance

b. Bearing Resistance

c. Net Section Resistance

d. Bolt Shear Resistance

Figure 1. Potential Limit States

4.4.3 Bearing on Connecting Part

Bearing resistance, the limit state of a local buckling of the part resulting from contact of the bolt on the connected part, is defined by the following equation:

$$F_{b,\,Rd} = 2.5\ f_u\ d\ t\ /\ \gamma_{M2}$$

where d = nominal bolt diameter, t = thickness of the thinner part or sheet.

If inadequate edge distance is provided, the bearing limit state may be preceded by longitudinal shear as defined as follows:

$$F_{b,Rd} = f_u\ e_1\ t\ /\ 1.2\ /\ \gamma_{M2}$$

4.4.4 Shear Resistance of Bolt

Although not common in cold-formed steel connections, bolt shear may limit the design resistance of a connection. The shear resistance of a bolt is prescribed by Eurocode3 as

$$F_{v,\,Rd} = 0.6\ f_{ub}\ A_s\ /\ \gamma_{M2} \text{ for strength grades 4.6, 5.6 and 8.8}$$

$$F_{v,\,Rd} = 0.5\ f_{ub}\ A_s\ /\ \gamma_{M2} \text{ for strength grades 4.8, 5.8, 6.8 and 10.8}$$

Where f_{ub} = ultimate tensile strength of the bolt and A_s = tensile stress area of a bolt. In lieu of calculations, the shear resistance is permitted to be determined by test in accordance with Section 9.

4.4.5 Tensile Resistance of Bolt

Bolt tensile strength is commonly not a design resistance limit, however calculations may be used to establish the tensile resistance using the following equation:

$$F_{t,\,Rd} = 0.9\ f_{ub}\ A_s\ /\ \gamma_{M2}$$

In lieu of calculations, the tensile resistance of a bolt may be determined by test in accordance with Section 9.

In addition to the tensile resistance of the bolt, a bolted connection in tension may experience a pull-through of the top sheet over the head of the bolt. This may be critical when the sheets are being connected. Eurocode 3 Part 1.3 does not prescribe a specific design resistance, but requires that the design resistance be defined by test. According to Part 1.3, pull-out of the fastener from the base material is not relevant for bolts.

4.4.6 Combined Tension and Shear

If the tension and shear resistance of a bolt have been determined by calculation, the interaction of a combined tension and shear loading may be verified by using:

$$F_{t,Sd}\ /\ F_{t,Rd} + F_{v,Sd}\ /\ F_{v,Rd} \leq 1.0$$

4.5 Screw Connections

Although bolts and welds are used in cold-formed steel construction, the predominant type of fastener employed in cold-formed steel construction is the screw. The choice of a screw for accomplishing the connection of cold-formed steel members is attributed to the ease of fabrication and installation (Figure 2 and 3).

Although there are similarities in the behavior and design of a bolt and a screw, specifying the appropriate screw is, however, more complex than specifying a bolt. Tapping screws are externally threaded fasteners that have the ability to "tap" their own internal mating threads when driven in the steel sheet. The more common tapping screws used in cold-formed steel construction are the self-drilling screw and the self-piercing screw.

The self-drilling screws are externally threaded fastener having the capability to drill their own hole and form, or tap, their own internal threads. The screws are manufactured from hardened steel and therefore the screw will not deforming their own thread and with our breaking during assembly. The self-drilling screw may be used in connections of multiple sheet thickness of 0.8 mm or thicker.

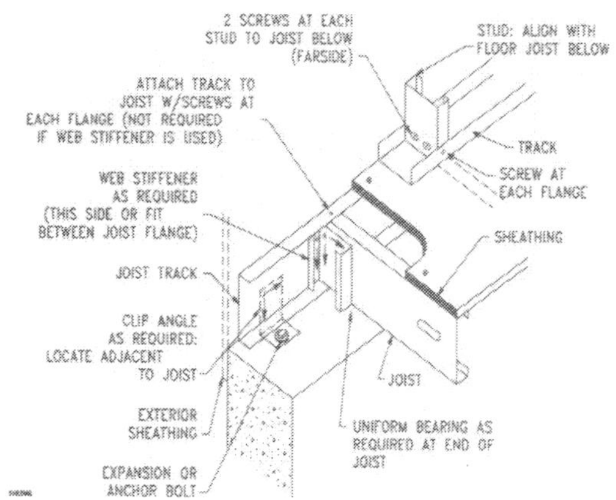

Figure 2. Common Self-Drilling Screw Applications

A self-piercing screw also is an externally threaded fastener, but the self-piercing screw has the capability of self-pierce thinner steel sheet. During the piercing process, sheet material forms a sleeve by extruding material and then the screw taps its own mating threads when driven. Similar to the self-drilling screw, the self-piercing screw is a one-side installation fastener. The self-piercing fastener is primarily used to attach rigid materials to thin steel sheet, for example attachment of wall sheathing to steel wall studs.

Specifying a strength (grade) of screw is not as easily accomplished as is the procedure for a bolt.

Eurocode 3, Part 1.3 lists six different strength grades of bolts that may be used in cold-formed steel, however, the manufacturer must define the strength of a screw.

If a tapping screw is used that requires a hole, the holes may be drilled or punched, although punched is the preferred for speed of fabrication. The diameter of a pre-drilled or punched hole for a screw should be in accordance with the manufacturers' guidelines. These guidelines should be based on

1. Applied torque should be higher than the threading torque.
2. Applied torque should be lower than thread stripping torque or head-shearing torque.
3. Thread torque should be smaller than 2/3 of the head-shearing torque.

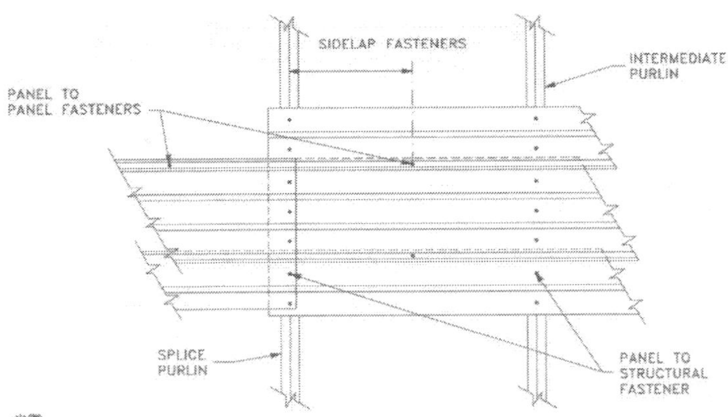

Figure 3. Common Self-Piercing Screw Applications

Proper installation of screws is important to achieve satisfactory performance. Power tools with adjustable torque controls and driving depth limitations are highly recommended.

When design is based on calculation, as summarized in Table 8.2 of Part 1.3. The design equations summarized by Table 8.2 are predicated on the thinnest sheet in the connection being next to the screw head. For connections transferring tension forces, the following limitations are affixed to the validity of the resistance equations:

$$0.5 \text{ mm} \leq t \leq 1.5 \text{ mm}$$
$$t_l \; 0.9 \text{ mm}$$

where t = thickness of the thinner sheet in the connection and t_l = thickness of the thicker sheet in the connection.

In addition to limiting thickness, the screw diameter has also prescribed limit. Generally, the screw diameter is limited to $3.0 \text{ mm} \leq d \leq 8.0 \text{ mm}$. These limits reflect the range of the screw diameters considered in the test programs. The design resistance equations are empirical and there limitations prescribed by the Eurocode 3 merely reflect the range of the tested parameters.

When subjected to shear forces, four general types of failure modes must be considered when designing a screw connection. When insufficient end distance from the centre of the fastener to the adjacent end of the connected part in the direction of the load, e_1, a longitudinal shear failure may result (Fig. 4a.). If sufficient end distance is provided, such that a longitudinal shear failure is prevented, bearing failure of the connected part may occur (Fig. 4b.). Although not a common limit state, fracture in the net section of the connected part must be evaluated (Fig. 4c.). Another potentially occurring limit state, shear failure of the screw, must also be investigated (Fig. 4d).

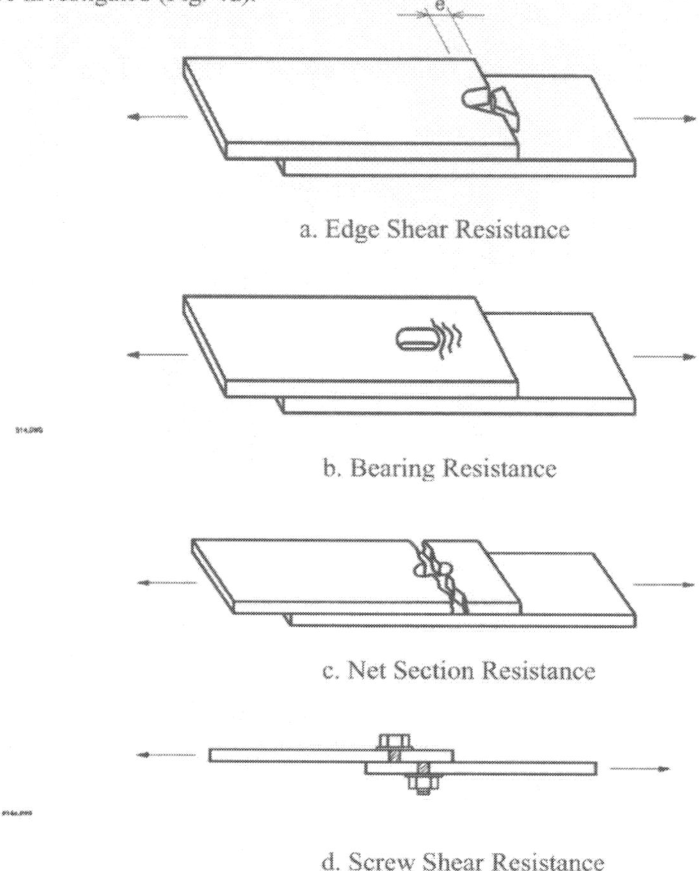

a. Edge Shear Resistance

b. Bearing Resistance

c. Net Section Resistance

d. Screw Shear Resistance

Figure 4. Potential Limit States

4.5.1 Longitudinal Shear

The minimum edge distance of each connected part, e_l, is not determined by a design equation, but in fact is address in the Eurocode3 by stipulating the minimum edge distance:

$$e_l \geq 3\ d$$

where e_l = end distance from the centre of the fastener to the adjacent end of the connected part and d = nominal diameter of the screw. This limit is a based on the performance of screw connections during laboratory testing. Tests have demonstrated that when the end distance is less than 3 d, longitudinal edge shear could occur.

4.5.2 Net Section Resistance

In Eurocode 3, Part 1.3, the design tension resistance on the net section of connected parts is based on the provisions of Section 8.4 where the net section resistance is defined by the following equation:

$$F_{n,Rd} = A_{net}\, f_u / \gamma_{M2}$$

4.5.3 Bearing on Connecting Part

Bearing resistance, the limit state of a local buckling of the part resulting from contact of the bolt on the connected part, is defined by the following equation:

$$F_{b,\,Rd} = \alpha\, f_u\, d\, t / \gamma_{M2}$$

where d = nominal bolt diameter, t = thickness of the thinner part or sheet, and α is defined as follows:

$$\text{if } t = t_l,\ \ \alpha = 3.2\,(t/d)^{0.5} \leq 2.1$$
$$\text{if } t_l \geq 2.5\,t,\ \alpha = 2.1$$
$$\text{if } t < t_l \leq 2.5\,t, \text{ obtain } \alpha \text{ by linear interpolation.}$$

The bearing resistance is function of the relationship between the thinner and thicker sheet thickness because for thicker sheet, i.e. if $t_l \geq 2.5$ t, the failure mode is a bearing type failure. However, for thinner sheet the failure mode is a tilting and tearing failure mode or a combination of tilting, tearing and bearing failure mode.

4.5.4 Shear Resistance of Screw

Screw shear resistance may limit the design resistance of a connection. Eurocode3 does not specifically prescribe the shear resistance of a screw. Thus, the shear resistance of a screw must be documented by testing in accordance with Section 9 Part 1.3.

To avoid a brittle and sudden fracture of a screw subjected to a shear force, the design resistance of the screw must be greater than the design resistance of the sheet as defined by the following equations:

$$F_{v,\,Rd} \geq 1.2\, F_{b,\,Rd}$$
$$F_{v,\,Rd} \geq 1.2\, F_{n,\,Rd}$$

4.5.5 Pull-Through Resistance of a Screw Connection

Screws are commonly used to connect thin sheathing to a structural member, for example a steel roof sheet attached to a roof purlin. When roof sheathing is subjected to a wind uplift load, a potential failure may be occur as the result of the sheathing fracturing around the head of the screw or the fracturing around the diameter of a washer. The sheathing pulls over the fastener head or washer, and the fastener remains engaged in the purlin flange. The design resistance for pull-through is defined by the following expressions:

$$F_{p,Rd} = d_w\, t\, f_u / \gamma_{M2}, \text{ for connections subject to static loads}$$
$$F_{p,Rd} = 0.5\, d_w\, t\, f_u / \gamma_{M2}, \text{ for connections subject to wind loads}$$

Where d_w = diameter of washer or diameter of fastener head.

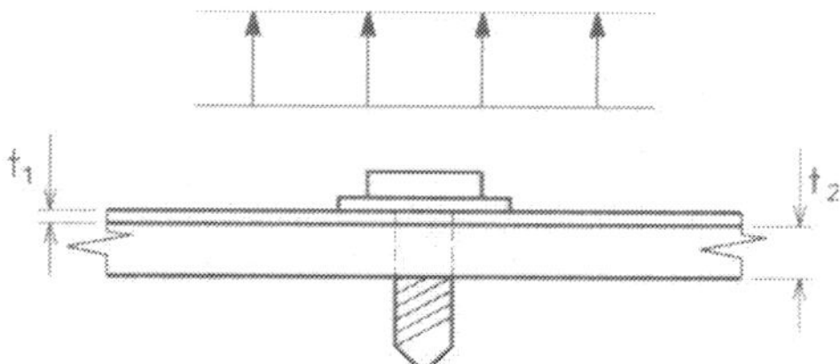

4.5.6 Pull-C

When a screw is subjected to a tension force and the sheet beneath the fastener head has sufficient strength such that the sheathing does not pull over the fastener head or washer, the screw may become disengaged from the supporting material, i.e. the flange of the purlin. In such cases, the design resistance for pull-out of the fastener from the supporting member is defined by the following equation:

$$F_{o,Rd} = 0.65\, d\, t_{sup}\, f_{u,sup} / \gamma_{M2}$$

Where t_{sup} = the thickness of the supporting member into which the screw is fixed, and $f_{u,sup}$ = the ultimate tensile strength of the supporting member.

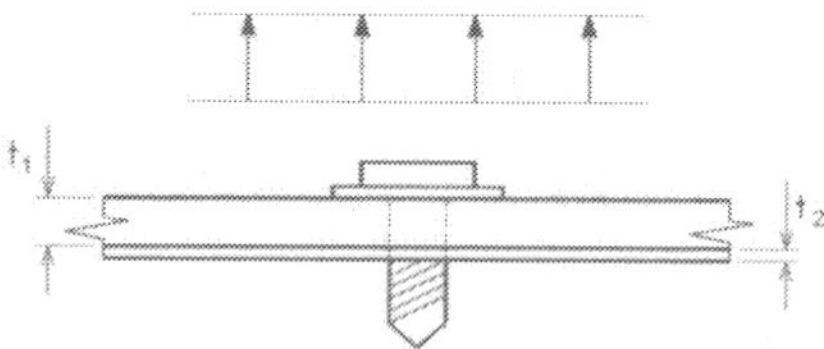

4.5.7 Tensile l

Screw tensile strength is commonly not a design resistance limit, however the tension resistance of a screw must be defined by test in accordance with Section 9.

To avoid a brittle and sudden fracture of a screw subjected to a tension force, the design resistance of the screw must be greater than the design resistance of the sheet as defined by the following equations:

$$F_{t,\,Rd} \geq n\, F_{p,\,Rd}$$
$$F_{t,\,Rd} \geq F_{o,\,Rd}$$

Where n = number of sheets that are fixed to the supporting member by the same screw.

4.5.8 Combined Tension and Shear

If the tension and shear resistance of a bolt have been determined by calculation, the interaction of a combined tension and shear loading may be verified by using:

$$F_{t,Sd} / F_{t,Rd} + F_{v,Sd} / F_{v,Rd} \leq 1.0$$

4.6 Welded Connections

4.6.1 Generals

Welds used to connect cold-formed steel may be either arc welds or resistance welds. In building construction, however, welds are generally made using the arc welding process. Resistance welds are commonly used for connecting thin sheet steels in the automotive or appliance. Arc welding is the process of fusing material together by an electric arc and the addition of weld filler metal.

Eurocode 3, Part 1.3 governs the design of welds for cold-formed steel construction. All welding consumables must conform to the requirements of ENV 1993-1-1. The Eurocode contains requirements for the following types of welds: (1) spot weld (2) arc seam weld, and (3) fillet welds. Both resistance welds and fusion arc spot welds are permitted.

The most common weld type to connect sheet-to-sheet or cross section-to-cross section is the fillet weld. Groove welds are commonly used during the roll forming process to connect flat sheet of one coil to the subsequent coil. Arc spot welds, commonly called puddle welds, are used extensively to attach deck and panels to bar joists or hot-rolled shapes.

Eurocode 3, Part 1.3 welding provisions provides guidelines for the following: workmanship, technique, and inspection.

4.6.2 Welding Processes

Welding electrodes should appropriately match the strength of the base metals. The following welding processes may be used for cold-formed steel construction: shielded metal arc welding (SMAW), gas metal arc welding (GMAW), flux cored arc welding (FCAW), gas tungsten arc welding (GTAW), and submerged arc welding (SAW).

SMAW

Shielded metal arc welding (SMAW), commonly known as stick electrode welding or manual welding, is the oldest of the arc welding processes. Versatility, simplicity and flexibility characterize it. The SMAW process commonly is used for tack welding, fabrication of miscellaneous components, and repair welding. SMAW has earned a reputation for depositing high quality welds dependably. It is, however, slower and more costly than other methods of welding, and is more dependent on operator skill for high quality welds.

FCAW

The flux cored arc welding process offers two distinct advantages over shielded metal arc welding. First, the electrode is continuous. This eliminates the built-in starts and stops that are inevitable with shielded metal arc welding. Not only does this have an economic advantage because the operating factor is raised, but the number of arc starts and stops, a potential source of weld discontinuities, is reduced. Another major advantage is that increased amperages can be used with flux cored arc welding, with a corresponding increase in deposition rate and productivity.

SAW

Submerged arc welding (SAW) differs from other arc welding processes in that a layer of fusible granular material called flux is used for shielding the arc and the molten metal. The arc is struck between the workpiece and a bare wire electrode, the tip of which is submerged in the flux. Since the arc is completely covered by the flux, it is not visible and the weld is made without the flash, spatter, and sparks that characterize the open-arc processes. The nature of the flux is such that very little smoke or visible fumes are released to the air.

GMAW

Gas metal arc welding (GMAW) utilizes equipment much like that used in flux cored arc welding. Indeed, the two processes are very similar, except GMAW uses a solid or metal cored electrode and leaves no appreciable amount of residual slag. Gas metal arc has not been a popular method of welding in the fabrication shop. A variety of shielding gases or gas mixtures may be used for GMAW. Carbon dioxide (CO_2) is the lowest cost gas, and while acceptable for welding carbon steel, the gas is not inert but active at elevated temperatures. This has given rise to the term MAG (metal active gas) for the process when (CO_2) is used, and MIG (metal inert gas) when predominantly argon-based mixtures are used.

GTAW

Gas tungsten arc welding (GTAW) or tungsten inert gas (TIG), as it is sometimes known, is a process where coalescence is produced by heating with an arc between a tungsten electrode and the base metal. The hot tungsten electrode, arc, and weld pool are shielded by an inert gas or mixture of inert gases. Filler metal may be added, if needed, by feeding a filler rod into the weld pool either manually or automatically.

4.6.3 Workmanship

ENV 1993-1-1 states that assembly and welding shall be carried out in such a way that the final dimensions are within the appropriate tolerances. In accordance with ENV 1993-1-1, the project specifications must stipulate details pertaining to welded connections that require special welding procedures, special levels of quality control, special inspection procedures, or special test procedures. In the United States, AWS D1.3 stipulates that the surfaces to be welded shall be smooth, uniform, and free of imperfections. When welding galvanized sheet, suitable ventilation shall be provided. Also, welding of sheet steels shall not be done when the ambient temperature is lower than -32° C (0° F); or when the surfaces are wet; or when the welder is exposed to inclement weather. The parts to be joined shall be brought into close contact to facilitate complete fusion. The closeness of the two parts can not be over-emphasized, especially for arc spot welds. If any gap exists between the members prior to spot welding, the strength of the weld may be substantially reduced. Also, to obtain consistently sound welds, the welding current must be controlled.

4.6.4 Inspection

ENV 1993-1-1 requires that the project specifications define special inspection procedures. In the United States, AWS D1.3 recognizes only visual inspection of welded sheet steel joints. The visual inspection shall determine compliance with contract documents. Particular emphasis shall be placed on verifying proper location, size, and length of a weld, in addition to the bead shape, reinforcement, and undercut.

4.6.5 Design Considerations

Of primary importance to the structural design engineer is the strength of a welded connection. Strength equations for the weld types are stipulated by the Eurocode 3, Part 1.3. The paramount difference between the strength of a welded connection in cold-formed steel and a welded connection in hot-rolled steel is the dominance of sheet tearing as a failure mode. Although the design provisions provide guidance on determination of the weld strength, the design is generally limited by the tearing of the base steel.

The design engineer must also consider workmanship, quality, and inspection when determining if a weld is an appropriate connection method.

A design equation is a mathematical relationship that models the failure of a welded connection. The following is an overview of the Standard's strength equations for the common weld types.

4.6.5.1 Design Equations for Spot Welds

Spot welds design resistance, as defined by Section 8.5 of Part 1.3, may be either resistance welded or fusion welded. However, in building construction spot welds are generally made using the arc welding process. Section 8.5 of Part 1.3 summarizes the strength design rules for arc spot welds. The primary focus of the design equations is the tearing of the sheet around the parameter of the weld, not the weld strength.

Spot welds may be used for connecting steel sheets having a parent thickness of less than or equal to 4 mm. However, a maximum thickness of 3 mm is stipulated for the thinner sheet. Additional limitations on the design resistance equations in Section 8.5 of Part 1.3 are as follows:

$$2\, d_s \leq e_1 \leq 6\, d_s$$

$$e_2 \leq 4\, d_s$$

$$3\, d_s \leq p_1 \leq 8\, d_s$$

$$3\, d_s \leq p_2 \leq 6\, d_s$$

where e_1 = end distance from the centre of the fastener to the adjacent end of the connected part, e_2 = edge distance from the centre of the fastener to the adjacent edge of the part in the direction perpendicular to the direction of the load transfer, p_1 = spacing of the centre-to-centre of the fasteners in the direction of the load transfer, p_2 = spacing of the centre-to-centre of the fasteners in the direction perpendicular to the direction of load transfer, and d_s = interface weld diameter.

Based on experimental measurements, the interface diameter is defined as follows:

For fusion welds: $d_s = 0.5\ t + 5$

For resistance welds: $d_s = 5\ (t)^{0.5}$

Note that when computing d_s the thickness must be in units of millimeters.

4.6.5.2 Tearing and Bearing Resistance

Bearing resistance, the limit state of a local buckling of the part resulting from contact of the bolt on the connected part, is defined by the following equation:

If $t \leq t_1 \leq 2.5\ t$:

$$F_{tb,\,Rd} = 2.7\ [(3)^{0.5}\ t\ d_s\ f_u\ /\ \gamma_{M2}]$$

If $t_1 > 2.5\ t$:

$$F_{tb,\,Rd} = 2.7\ [(3)^{0.5}\ t\ d_s\ f_u\ /\ \gamma_{M2}]$$

$$F_{tb,\,Rd} \leq 0.7\ t\ d_s\ f_u\ /\ \gamma_{M2}$$

$$F_{tb,\,Rd} \leq 3.1\ t\ d_s\ f_u\ /\ \gamma_{M2}$$

where t = thickness of the thinner connected part, t_1 = thickness of the thicker connected part, f_u = ultimate tensile strength of the part and γ_{M2} = 1.25. The weld diameter, $d_s = 0.5\ t + 5$ at the interface of the two parts being connected.

4.6.5.3 End Resistance

Similar to a bolt or screw connection, a spot weld may fail in an edge shear failure mode. The end resistance for a spot weld is defined by the following relationship:

$$F_{e,\,Rd} = 1.4\ t\ e_1\ f_u\ /\ \gamma_{M2}$$

4.6.5.4 Net Section Resistance

The connected part, when subject to an in-plane tension force, must be investigated for the potential of a ductile fracture in the net section. Section 8.5, Part 1.3 prescribes the following net section resistance equation:

$$F_{n,Rd} = A_{net}\ f_u\ /\ \gamma_{M2}$$

where A_{net} = net tension area.

4.6.5.5 Shear Resistance

When a fusion or resistance spot weld is formed, no weld metal is added to accomplish the connection. Thus, the shear strength, or design resistance, of a spot weld is a function of the ultimate tensile strength of the connected part as determined by the following equation:

$$F_{v,Rd} = 0.25\ \pi\ d_s^2\ f_u / \gamma_{M2}$$

To ensure that the connection has adequate ductility, the following additional conditions are stipulated in Section 8.5:

$$F_{v,Rd} \geq 1.25\ F_{tb,Rd} \text{ and } F_{v,Rd} \geq 1.25\ F_{e,Rd} \text{ and } F_{v,Rd} \geq 1.25\ F_{n,Rd}$$

4.6.5.6 Design Equations for Lap Welds

In building construction, however, welds are generally made using the arc welding process. Section 8.6 of Part 1.3 of the Eurocode 3 defines the design resistance for the arc spot weld (puddle weld), the arc seam weld, and the fillet weld. However, the use of Section 8.6 of Part 1.3 is limited to welded connections where the parent material, that is connected part, is less than or equal to 4 mm. For welding of thicker material, the design resistance must be defined by ENV 1993-1-1.

When designing a welded connection, the weld size must be chosen such that the resistance of the connection is governed by the thickness of the connected part, not the weld itself. This requirement may be assumed to be satisfied if the throat size of the weld is at least equal to the thickness of the connected part.

4.6.5.7 Fillet Weld Resistance

When fillet welds are used in cold-formed steel construction, the fillet weld throat is commonly at least equal to the thickness of the connected part or sheet. Based on research findings, the ultimate strength of a fillet weld connection has been found to occur by tearing the sheet, not failure of the weld (Figure 7). In most cases, the higher strength of the weld material prevents a weld failure. The resistance equations in Section 8.6.2 of Part 1.3 are based on sheet tearing.

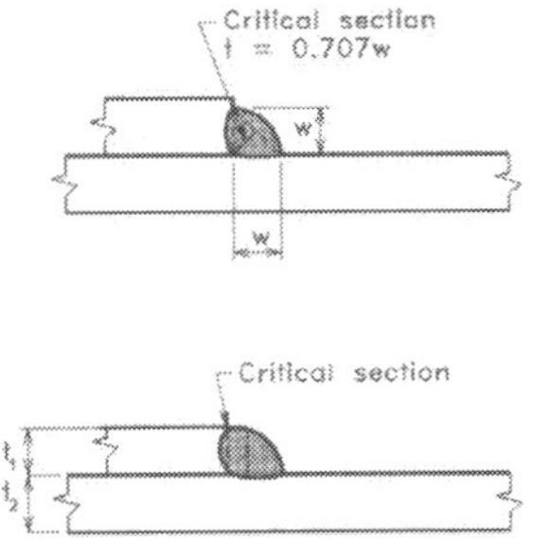

Figure 7. Fillet Weld Failure Modes

Because sheet tearing is the governing failure mode, the direction of the load with respect to the axis of the weld, will influence the connection resistance. Section 8.6.2 of Part 1.3 contains the following equations for determining the design resistance of a fillet weld:

> For a side fillet weld that comprises of a pair of side fillets, that is a fillet weld parallel to the direction of the applied force,
>
> $$F_{w,Rd} = t\, L_{w,s}\, (0.9 - 0.45\, L_{w,s} / b)\, f_u / \gamma_{M2}$$
>
> For an end fillet weld, that is a fillet weld perpendicular to the direction of the applied force,
>
> $$F_{w,Rd} = t\, L_{w,e}\, (1 - 0.3\, L_{w,e} / b)\, f_u / \gamma_{M2}$$

where b = width of the connected part or sheet, $L_{w,e}$ = effective length of the end fillet weld, $L_{w,s}$ = effective length of a side fillet weld.

When a combination of a side fillets and end fillets are used to resist the same applied force, the total resistance is taken as the sum of the resistance of the side and end fillets. The effective length is taken as the overall length of a full-size fillet weld including end returns. If the weld size is full-size along the entire length, no reduction in the effective length need be made for either the start or termination of the weld.

Short welds are not effective to transfer load, therefore Section 8.6.2 of Part 1.3 indicates that fillet welds with effective lengths less than 8 times the thickness of the thinner connected part should not be designed to transmit forces.

4.6.5.8 Arc Spot Weld Resistance

Eurocode 3 Part 1.3 only contains design resistance equations for connections that transmit shear forces (Figure 8). Arc spot welded connections are often made by melting through the top sheet(s) and fusing the sheets together with additional filler metal. Thus, the spot welds should not be used through connected parts or sheets with a total thickness of more than 4 mm or the thinnest connected part is more than 4 mm thick. To ensure proper penetration and to avoid excessive burning of the sheet, if the thickness of the sheet is less than 0.7 mm, a weld washer must be used.

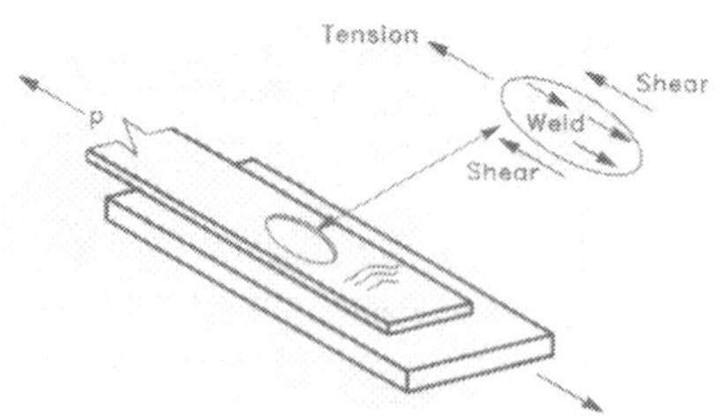

Figure 8. Arc Spot Weld Behavior

4.6.5.8.1 Circular Arc Spot Weld. When determining the design resistance of a circular arc spot weld both the weld resistance and the sheet resistance must be evaluated. The failure mode of the sheet may be either a tearing of the sheet along the contour of the weld in a shear, tension, or combined shear and tension failure plane (Figure 2). Sheet tearing combined with a buckling near the trailing edge of the weld may also occur. The weld metal may fail in a shear failure of in the fused area.

The design shear resistance of a circular arc spot weld is determined as follows:

$$F_{w,Rd} = (\pi/4)\, d_s^2 \text{ x } 0.5\, f_{uw} / \gamma_{M2}$$

where f_{uw} = ultimate tensile strength of weld electrode and d_s = the interface diameter of the spot weld. The interface diameter is the diameter of the weld is the fusion diameter of the weld and is defined by the following equation based on measured fusion diameters:

$$d_s = 0.7\, d_w - 1.5\, t$$

where d_w = visible diameter of the arc spot weld and t = thickness of the connected sheet(s).

The above equation for $F_{w,Rd}$ assesses the shear resistance of the weld itself, to evaluate the design resistance of the connected sheet(s), the following equations are provided in Section 8.6.3 of Part 1.3:

If $d_p/t \le 24\,\varepsilon$:

$$F_{w,Rd} = 1.33\, d_p\, t\, f_u / \gamma_{M2}$$

If $24\,\varepsilon < d_p/t \le 41.5\,\varepsilon$:

$$F_{w,Rd} = 0.17\,(d_p + 164\,\varepsilon\, t)\, t\, f_u / \gamma_{M2}$$

If $d_p/t \ge 41.5\,\varepsilon$:

$$F_{w,Rd} = 0.84\, d_p\, t\, f_u / \gamma_{M2}$$

where t = thickness of the sheet(s) connected (Figure 9) and d_p = effective peripheral diameter of weld. The effective peripheral weld is essentially an average weld diameter through the thickness of the sheet(s) being welded and is defined by the following equations:

Figure 9. Arc Spot Weld

4.6.5.8.2 Elongated Arc Spot Weld. The behavior of an elongated arc seam weld is similar to the behavior of a circular arc spot weld. Section 8.6.3 of Part 1.3 prescribes a design equation for the weld strength but typically the strength of an arc seam weld is governed by tearing of the sheet at the perimeter of the weld.

The design shear resistance of a elongated arc spot weld is determined as follows:

$$F_{w,Rd} = [(\pi/4)\, d_s^2 + L_w\,] \times 0.5\, f_{uw} / \gamma_{M2}$$

where f_{uw} = ultimate tensile strength of weld electrode, L_w = length of the elongated arc spot weld, and d_s = the interface diameter of the spot weld. The interface diameter is the diameter of the weld is the fusion diameter of the weld and is defined by the following equation based on measured fusion diameters:

$$d_s = 0.7\, d_w - 1.5\, t$$

where d_w = visible diameter of the arc spot weld and t = thickness of the connected sheet(s).

The above equation for $F_{w,Rd}$ assesses the shear resistance of the weld itself, to evaluate the design resistance of the connected sheet(s), the following equations are provided in Section 8.6.3 of Part 1.3:

$$F_{w,Rd} = (0.4\, L_w + 1.33\, d_p)\, t\, f_u / \gamma_{M2}$$

where t = thickness of the connected sheet(s).

Arc welding is a safe occupation when sufficient measures are taken to protect the welder from potential hazards. When these measures are overlooked or ignored, welders can encounter such dangers as electric shock, over-exposure to radiation, fumes and gases, and fire and explosion; any of these can result in fatal injuries. Everyone associated with the welding operation should be aware of the potential hazards and ensure that safe practices are employed.

4.7 Shear Lag

When a tension member is not connected through all of the elements of the cross section, the stress distribution in the cross section is nonuniform. An example of the nonuniform stress distribution is when an angle is connected through only one leg,. This phenomenon is called "shear lag". Shear lag has a weakening effect on the cross section because the connected element of the cross section becomes overloaded and the unconnected element is not fully stressed.

Shear lag occurs in both bolted and welded connections. Eurocode 3 does not provide adequate design guidance for shear lag. Thus, this paper will summarize the United States design rules for shear lag. These rules were first adopted in the 1999 Supplement to the *Specification for the Design of Cold-Formed Steel Structural Members* (1996) and appear in the subsequent edition of the specification (North, 2001).

4.7.1 Bolted Connections

In Eurocode 3, Part 1.3, the design tension resistance on the net section of connected parts is based on the provisions of Sections C5.2 and 8.4 whichever is smaller.

Section C5.2 prescribes the limit state of yielding in the gross section as

$$N_{t,Rd} = f_{ya} A_g / \gamma_{MO}$$

where f_{ya} = average yield strength, A_g = gross area of the cross section, and γ_{MO} =1.10. In addition to the limit state on yielding in the gross section, Section C5.2 also limits $N_{t,Rd}$ to the fracture in the net section resistance as prescribed by Section 8.4.

Net section resistance is defined by the following equation:

$$F_{n,Rd} = (1+3r(d_o/u - 0.3))A_{net}\, f_u/\, \gamma_{M2} \leq A_{net}\, f_u / \gamma_{M2}$$

where r = number of bolts at the cross section / total number of bolts in the connections, and u = 2 e_2 with a limit of $u \leq p_2$, the spacing centre-to-centre of fasteners in the direction perpendicular to the direction of load.

The equation for $F_{n,Rd}$ is based on tests of flat sheet connections. Thus, the quantity in the parentheses is a reduction factor to reflect that the stress is not uniform in the connection cross section. This non-uniform stress is called shear lag. Shear lag is more critical in cross sections having multiple elements, such as angles and C-sections.

Eurocode 3 does not contain design rules for addressing shear lag in cross sections. For guidance the designer may refer to the American Iron and Steel Institutes specification (North, 2001; Specification, 1996). For other than flat sheet connections, The American Iron and Steel Institute's specification states that the nominal strength, P_n, is to be determined as follows:

$$P_n = A_e F_u$$

in which F_u = tensile strength of the connected part and A_e = effective net area. The use of an effective or reduced area concept reflects the reduction in cross-section capacity that is the result of the non-uniform stress distribution.

The effective area is determined as follows:

$$A_e = U A_n$$

where A_n = net area, and U is the shear lag reduction factor defined as follows:

(a) For angle members having two or more bolts in the line of force:

$$U = 1.0 - 1.20\, \bar{x} / L < 0.9$$

but U shall not be less than 0.4

(b) For channel members having two or more bolts in the line of force:

$$U = 1.0 - 0.36\ \bar{x}/L < 0.9$$

but U shall not be less than 0.5

where $\bar{x}$ = distance from the shear plane to the centroid of the cross section and L = length of connection

U is to be taken as 1.0 for members when the load is transmitted directly to all of the cross-section elements (Figure 10).

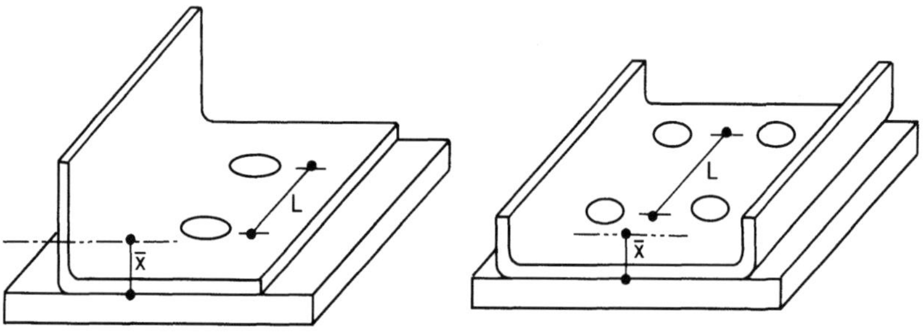

Figure 10. Bolted Connection Shear Lag Parameters

4.7.2 Welded Connections

When utilizing welded connections, if the tension member is not connected through all of the cross-section elements, shear lag will have a weakening effect on the tension capacity of the member. The 1999 Supplement to the *Specification for the Design of Cold-Formed Steel Structural Members* (1996) and the 2001 edition of the specification (North, 2001) stipulates the following design guidelines:

$$P_n = A_e F_u$$

in which F_u is the tensile strength of the connected part and A_e is the effective area.

The effective area, A_e = UA, depends to the geometry of the welded connection and the application of the load as follows:

1. When the load is transmitted only by transverse welds, A = area of directly connected elements and U = 1.0

2. When the load is transmitted only by longitudinal welds or by longitudinal welds in combination with transverse welds, A = gross area of member, A_g. When the load is transmitted directly to all of the cross-section elements U = 1.0, otherwise the reduction factor is determined as follows:

(a) For angle members:

$$U = 1.0 - 1.20\ \bar{x}/L < 0.9$$

but U shall not be less than 0.4

(b) For channel members:

$$U = 1.0 - 0.36\ \bar{x}/L < 0.9$$

but U shall not be less than 0.5

where $\bar{x}$ = distance from shear plane to centroid of the cross section, and L = length of longitudinal welds (Figure 11).

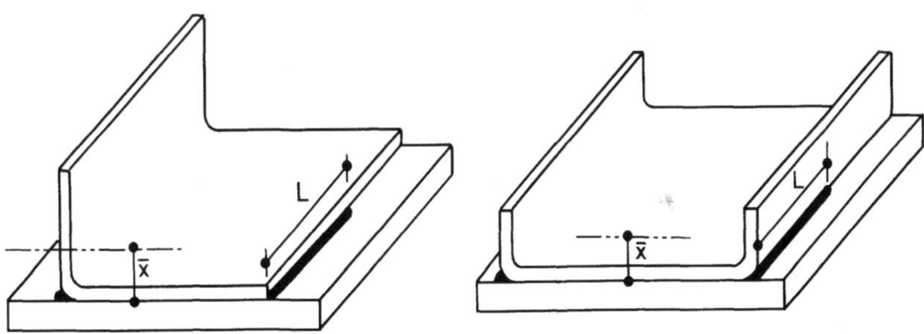

Figure 11. Welded Connection Shear Lag Parameters

4.8 Fatigue and Cyclic Loading

4.8.1 Introduction

Fatigue in a structural steel member or connection is the process of initiation and subsequent growth of a crack under the action of a cyclic or repetitive load. The fatigue process commonly occurs at a stress level less than the static failure condition.

When fatigue is a design consideration, its' severity is determined primarily by three factors: (1) the number of cycles of loading, (2) the type of member and connection detail, and (3) the stress range at the detail under consideration (Fisher et al. 1998).

Although fatigue design guidelines have existed for hot-rolled steel structural members and connections there have been no generally accepted design guidelines for addressing fatigue in a cold-formed steel member or connection.

The fatigue design recommendations provided in the AISI Specification (North, 2001) are based on a review of available test data. No additional experimental studies were performed to support the suggested design recommendations.

Research by Barsom (1980) and Klippstein (1988, 1985, 1981, 1980) developed fatigue information on the behavior of sheet and plate steel weldments and mechanical connections. Using regression analysis, mean fatigue life curves (S-N curves) with the corresponding standard deviation were developed. The fatigue resistance S-N curve has been expressed as an exponential relationship between stress range and life cycle (Fisher, 1970). The general relationship is often plotted as a linear log-log function,

$$\log N = C_f - m \log F_{SR}$$

$$C_f = b - (n * s)$$

where N = number of full stress cycles;
m = slope of the mean fatigue analysis curve;
F_{SR} = effective stress range;
b = intercept of the mean fatigue analysis curve from Table C1;
n = number of standard deviations to obtain a desired confidence level
= 2 for C_f given in the Specification;
s = approximate standard deviation of the fatigue data
= 0.25 (Klippstein, 1988).

The data base for these design provisions are based upon cyclic testing of real joints; therefore, stress concentrations have been accounted for by the category in Table 1 of the Specification.

4.8.2 Design Specification

When cyclic loading is a design consideration, the provisions of the AISI specification (North, 2001) apply to stresses calculated on the basis of unfactored loads. The maximum permitted tensile stress due to unfactored loads is to be taken as 0.6 F_y.

Stress range is defined as the magnitude of the change in stress due to the application or removal of the unfactored live load. In the case of a stress reversal, the stress range shall be computed as the sum of the absolute values of maximum repeated tensile and compressive stresses or the sum of the absolute value of the maximum shearing stresses of opposite direction at the point of probable crack initiation.

The occurrence of full design wind or earthquake loads is too infrequent to warrant consideration of fatigue design. If the live load stress range is less than the threshold stress range, F_{TH}, given in Table 1, evaluation of fatigue resistance is not also required.

Evaluation of fatigue resistance is also not required if the number of cycles of application of live load is less than 20,000.

The calculated stresses shall be based upon elastic analysis. Stresses shall not be amplified by stress concentration factors. For bolts and threaded rods subject to axial tension, the calculated stresses shall include the effects of prying action. For combined axial and bending stress conditions, the maximum stresses shall be those determined for concurrent arrangements of applied load. For members having symmetric cross sections, the fasteners and welds shall be arranged symmetrically about the axis of the member, or the total stresses including those due to eccentricity shall be included in the calculation of the stress range. For axially stressed angle members where the center of gravity of the connecting welds lies between the line of the center of gravity of the angle cross section and the center of the connected leg, the effects of eccentricity shall be ignored. However, if the center of gravity of the connecting welds lies outside this zone, the total stresses, including those due to joint eccentricity, shall be included in the calculation of the stress range.

The range of stress at service loads shall not exceed the design stress range computed by the following equation:

$$F_{SR} = (C_f/N)^{0.333} \geq F_{TH}$$

where F_{SR} = design stress range, C_f = constant from Table 1, N = number of stress range fluctuations in the design life, F_{TH} = threshold fatigue stress range given by Table 1.

Table 1. Fatigue Design Parameters for Cold-Formed Steel Structures

Description	Stress Category	Constant C_f	Threshold F_{TH} (MPa)	Illustrative Example
As received base metal and components with as rolled surfaces, including sheared edges and cold-formed corners.	I	$3.2\text{x}10^{10}$	172	Fig. 12
As received base metal and weld metal in members connected by continuous longitudinal welds.	II	$1.0\text{x}10^{10}$	103	Fig. 13
Welded attachments to a plate or a beam, transverse fillet welds, and continuous longitudinal fillet welds less than and equal to 51 mm. Bolt and screw connections and spot welds.	III	$3.2\text{x}10^{9}$	110	Fig. 14, 15
Longitudinal fillet welded attachments greater than 51 mm parallel to the direction of the applied stress, and intermittent welds parallel to the direction of the applied force.	IV	$1.0\text{x}10^{9}$	62	Fig. 14

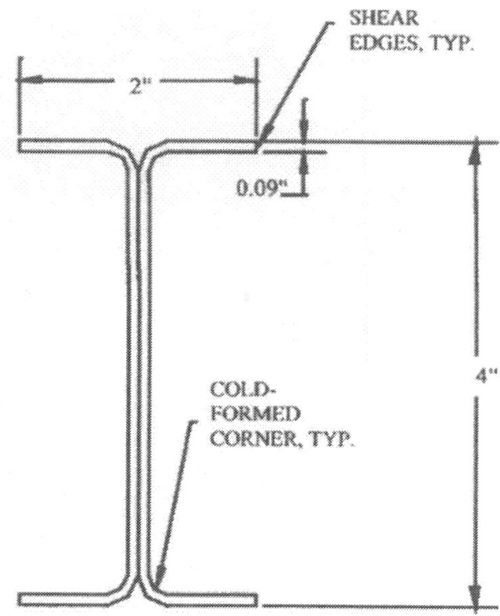

Figure 12. Typical detail for category I

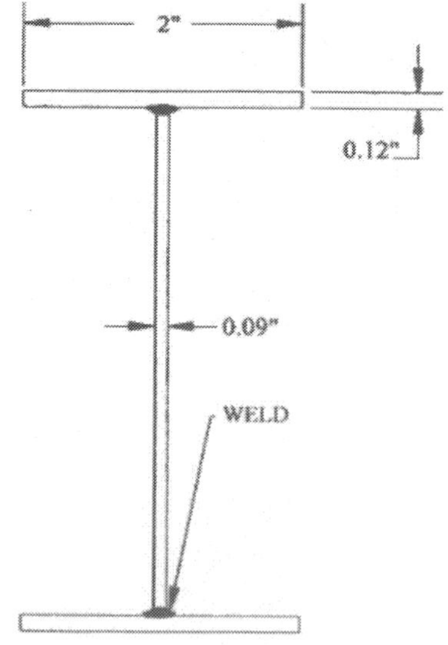

Figure 13. Typical Detail for Category II

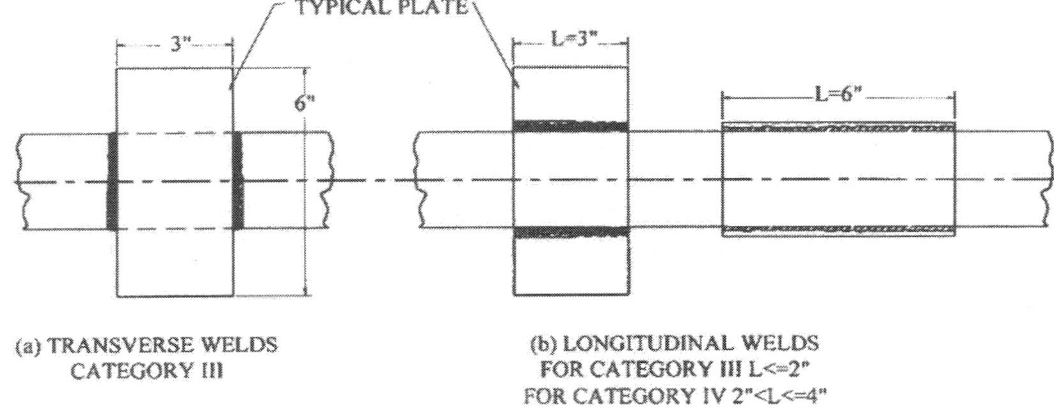

Figure 14. Typical Attachments for Categories III and IV

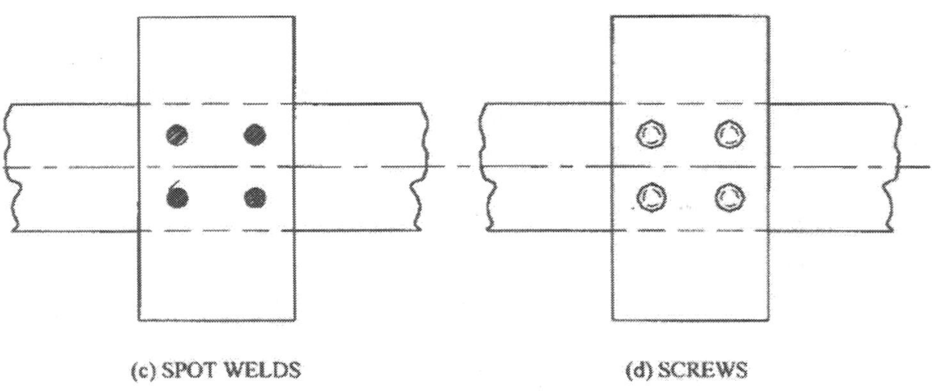

Figure 15. Typical Attachments for Category III

4.9 Special Connection technique fir cold-formed steel thin materials

4.9.1 Clinch Fastening

The clinching process has been used for decades in the automotive and appliance industries. Unfortunately, the application of clinching in the construction industry was hampered by tooling requirements. However, in recent years hand held tools have been developed to facilitate the fabrication of light steel frame structures. Both in-plant and job site tools are now available. The clinch can be used for the fabrication of floors, walls and roof members. To utilize a clinch connection the joint must be accessible from both sides and the predominant forces or load on the joint should be in shear.

Extensive testing, both in Europe and the United States, has been conducted for clinch connections. The test programs have explored both shear and tension load applications, as well as cyclic loading, and end and edge distance requirements. These test programs have discovered that the load carrying capacity of a clinch is proportional to the clinch geometry. The larger the clinch size, the greater is the load carrying capacity. Both round and rectangular clinches have been studied. The tensile strength and thickness of the sheets being joined also influence the structural performance. Tests have shown that coating will not affect the strength of a clinch. However, in connections were two unequal sheets are connected together a greater strength may be achieved when the thicker sheet is on the punch side.

Regardless of the clinch geometry, round or rectangular, the joint is formed by punch pressing metal into a die with sufficient force to cause metal to flow sideways forming a lock with the bottom sheet (Figure 16).

The strength of a rectangular clinch is orientation dependent because the shear strength is greater perpendicular to the long side of the clinch. The strength of a round clinch is not orientation dependent.

Research has shown that to achieve good quality connections, the following conditions should be considered:

1. Clinches should be placed greater than 1.5 punch diameters from the edge of the sheet.
2. In stud-to-track connections, the clinch must be sufficient distance from the web-to-flange or lip-to-flange radii to enable the die to seat squarely on the surface of the sheet.
3. The sheet ductility and thickness must be within the operating parameters of the equipment.
4. The tooling must be aligned to the proper settings.
5. The tooling must be properly maintained.

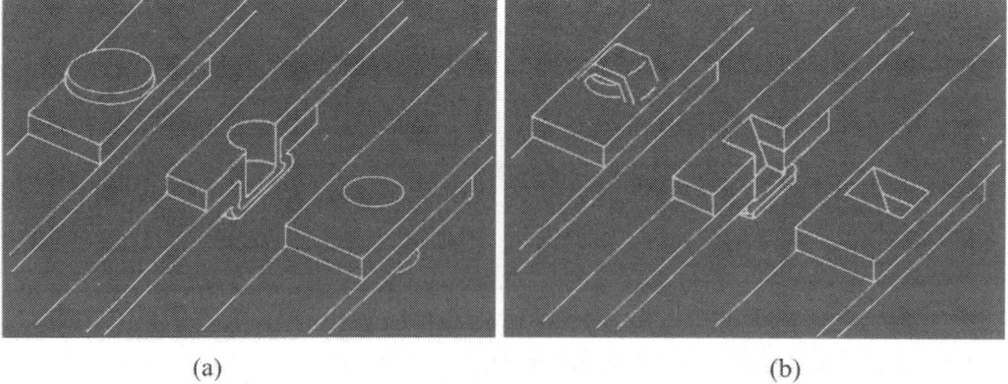

(a) (b)

Figure 16. Square and Round Clinch, top view, bottom view and cross-section view

4.9.2 Rosette Fastening

The Rosette connector is also a relatively new automated approach for fabricating cold-formed steel components. Typical applications are in the fabrication of stud wall systems and roof truss systems.

The Rosette connection is made by first punching holes in one part of the joint and forming collared holes in the other part to be joined. Next, the collars are snapped into the holes during loose assembly of each joint. Finally, the Rosette tool head penetrates the hole at each joint, the head expands and is then pulled back by hydraulic force. The expanded tool head crimps the collar part against the hole part completing the connection. The joining process is illustrated in Figure 17 and the finished Rosette connection is shown in Figure 18.

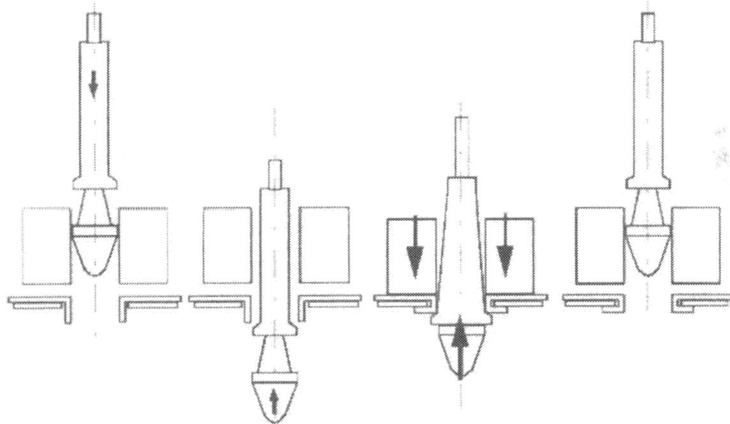

Figure 17. Rosette Joint Fabrication

Figure 18. Completed Rosette Joint

The resistance of the connection is dependent upon the thickness of the connected sheets and must be determined by test in accordance with Section 9.

4.9.3 Cold-Formed Seams

The behavior and subsequently the design of a roof system is primarily dependent on the type of roof panel. There are two general categories of roof panels, the conventional through-fastened panel, and the standing seam panel. The through-fastened panel, although it has its place in the marketplace, is rapidly being replaced by the standing seam panel.

Standing seam roof systems were first introduced in the late 1960's, and today many manufacturers produce standing seam panels. A difference between the standing seam roof and through-fastener roof is in the manner in which two panels are joined to each other. The seam between two panels is made in the field with a tool that makes a cold formed weather-tight joint. The joint is made at the top of the panel. The standing seam roof is also unique in the manner in which it is attached to the purlins. The attachment is made with a clip concealed inside the seam. This clip secures the panel to the purlin and may allow the panel to move when experiencing thermal expansion or contraction (Figure 19).

A continuous single skin membrane results after the seam is made since through-the-roof fasteners have been eliminated. The elevated seam and single skin member provides a weather tight system. The ability of the roof to experience unrestrained thermal movement eliminates damage to insulation and structure (caused by temperature effects which built-up and through fastened roofs commonly experience). Due to the superiority of the standing seam roof, most manufacturers are willing to offer considerably longer guarantees than those offered on lap seam roofs.

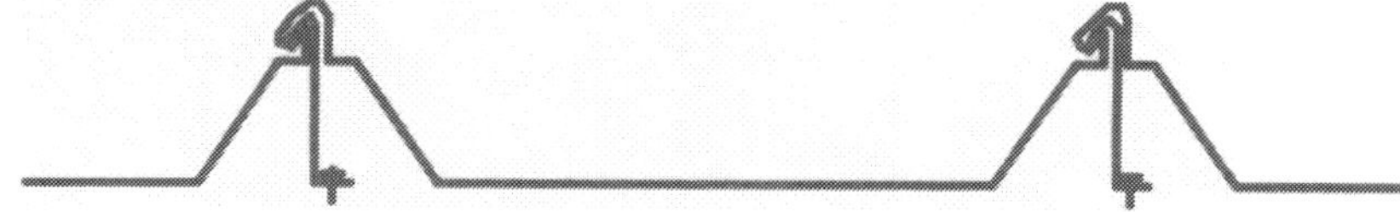

Figure 19. Standing Seam Roof Panel

The design strength of the standing seam panel, especially for a wind uplift loading condition must be determined by test.

4.9.4 Self-Piercing Pin

The use of pneumatically driven pins has been common for about 15 years in the United States. Pins used for cold-formed steel framing are smaller in diameter to pins used for thicker steel. The pins are installed by using a hand held pneumatic tool.

Pneumatic driven pins are proprietary products that have unique characteristics and performance capabilities that vary with the manufacturer. Variations can be found in the head size, shank geometry, and the length, diameter and point of the pin. In the United States, pin manufacturers publish design values for their product (Figure 20).

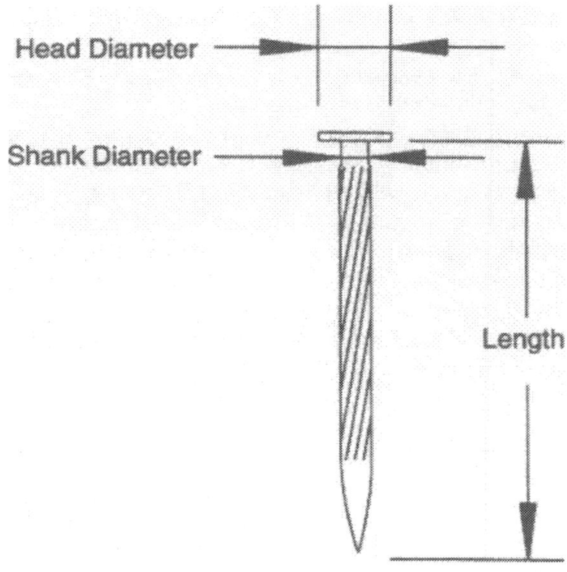

Figure 20. Self-Piercing Nail

References

Barsom, J. M., Klippstein, K. H., and Shoemaker, A. K. (1980), "Fatigue Behavior of Sheet Steels for Automotive Applications," Research Report SG 80-2, American Iron and Steel Institute

Cold-Formed Steel Framing Design Guide (2002), Design Guide CF02-1, American Iron and Steel Institute, Washington, D.C.

Commentary on the Specification for the Design of Cold-Formed Steel Structural Members (1996), American Iron and Steel Institute Washington, D.C.

Fisher, J. W., Kulak, G. L., and Smith, I. F.C. (1998), "A Fatigue Primer for Structural Engineers," National Steel Bridge Alliance.

Fisher, J. W. , Frank, K. H., Hirt, M. A., and McNamee, B. M. (1970), "Effect of Weldments on the Fatigue Strength of Beams," National Cooperative Highway Research Program, Report 102, Washington, D.C.

Klippstein, K. H. (1980), "Fatigue Behavior of Sheet Steel Fabrication Details," Proceedings of the Fifth International Specialty Conference on Cold-Formed Steel Structures, University of Missouri-Rolla

Klippstein, K. H. (1981), "Fatigue Behavior of Steel-Sheet Fabrication Details," SAE Technical Paper Series 810436, International Congress and Exposition, Detroit, MI

Klippstein, K. H. (1985), "Fatigue of Fabricated Steel-Sheet Details - Phase II," SAE Technical Paper Series 850366, International Congress and Exposition, Detroit, MI

Klippstein, K. H. (1988), "Fatigue Design Curves for Structural Fabrication Details Made of Sheet and Plate Steel," unpublished AISI research report.

Light Guage Steel Engineers Association (1999), "Integral (Clinched) Fastening of Cold-Formed Steel," Tech. Note 560c, Washington, D.C.

Light Guage Steel Engineers Association (1998), "Pneumatically Driven Pins for Wood Based Panel Attachment," Tech. Note 561b, Washington, D.C.

North American Specification for the Design of Cold-Formed Steel Structural Members (2001), American Iron and Steel Institute, Washington, D.C.

Makelainen, P., Kosti, J. Kaitila, O., and Sahramaa, K.J. (1998), " Study of Light-Guage Steel Roof Trusses with Rosette Connections," Proceedings of the 14th International Specialty Conference on Cold-Formed Steel Structures," University of Missouri-Rolla

North American Specification for the Design of Cold-Formed Steel Structural Members (2001), American Iron and Steel Institute, Washington, D.C.

Specification for the Design of Cold-Formed Steel Structural Members (1996), American Iron and Steel Institute Washington, D.C.

Chapter 5: Stainless Steel Structures

K.J.R. Rasmussen

Department of Civil Engineering, University of Sydney, Sydney, Australia
E-mail: K.Rasmussen@civil.usyd.edu.au

5.1 Introduction

Stainless steel structures are used in a wide range of applications, including plane and three-dimensional trusses, mullions in facade construction, canopies, roof sheeting, silos, portal framed overhead wiring structures (for railway services), and general construction in chemical, marine and other corrosive environments. Stainless steel is also used as concrete rebars and fixings for brickwork.

The tonnage price of stainless steel is typically 5-7 times that of carbon steel, depending on the type of alloy. Hence, stainless steel is likely to be used in specialised construction to take advantage of the superior material properties, (mainly to resist corrosion or maintain strength at elevated temperatures), reduce maintenance costs, and/or provide aesthetically pleasing appearances. In any case, the selection of the appropriate stainless steel alloy is important. Unlike construction in conventional carbon steel[1] where only two or three grades of steel would be considered, a large number of stainless steel alloys have been developed to provide different corrosion resistances, hardnesses, mechanical properties, weldabilities, etc. These alloys have different mechanical properties which needs to be recognised in their structural design. Furthermore, the mechanical properties of stainless steel alloys differ from those of carbon steels and so stainless steel structures should not simply be designed using specifications for carbon steel structures.

An introductory guide to the material selection and a summary of the mechanical properties pertinent to structural design are presented in Sections 5.2 and 5.3. Much information is available on these subjects, including the guidelines (NiDI 1990; AISI 1991a) distributed by the Nickel Development Institute (see http://www.nidi.org), and a paper by Mann (1993). Section 5.4 explains how the nonlinear stress-stain curves characterizing stainless steel alloys affect the structural design of members and plate elements. Sections 5.5 and 5.6 describe the Australian, American and European design provisions for stainless steel columns and plate elements. Section 5.7 summarises the Australian, American and European design provisions for connections in stainless steel, including bolts, welds and welded tubular joints. Worked examples are included on the design of columns and bolted connections.

---

[1] Throughout this chapter, the term 'carbon steel' refers to low alloy construction steels covered by structural design standards, such as Eurocode3, Part 1.1, the AISC-LRFD Specification and the Australian steel structures standard AS4100.

5.2 Material selection

5.2.1 Background

Stainless steel is a generic term covering a large group of iron-based alloys. Stainless steel is characterised by having chromium content greater than 12%, (although there is some contention about this percentage which is found varying between 10 and 12% in the literature).

The basic corrosion resisting mechanism of stainless steel is that when exposed to air, chromium combines with oxygen to form a thin, non-porous layer of chromium-oxide which protects the iron against corrosion. This layer, known as the "passive-film"', is highly resistant against chemical attacks and is capable of regenerating itself if disturbed, provided oxygen is present. The passive film requires about 11% chromium to form. Increasing the chromium content and adding other alloys, notably nickel and molybdenum can enhance the stability of the layer.

Much information is available on the metallurgy of stainless steel. It is a wide subject because the large number of alloys that have been developed for a diversity of applications (most of them non-structural) all have different metallurgical properties. An introductory text suitable for structural engineers can be found in Lula (1965). Useful background information is also available from the American Iron and Steel Institute (AISI 1991a) the Australian Stainless Steel Development Association (ASSDA 2002), as well as the appendices of Eurocode3, Part 1.4 (Eurocode3-1.4 1996) and the Australian Standard for cold-formed stainless steel structures (AS/NZS4673 2001). The American Iron and Steel Institute publication (AISI 1974b) presents a concise description of the most common alloys, detailing their chemical compositions and mechanical properties.

5.2.2 Main stainless steel groups and identification

The stainless steel alloys used for structural applications are divided into three main groups: austenitic, ferritic, and duplex, distinguished mainly by differences between their atomic structure (Lula 1965; AISI 1991a). While in the past the austenitic group has been used predominantly, interest is now growing for the duplex group. The three groups are summarised as follows:

Austenitic These alloys have face-centered atomic structure. They can be hardened by cold-working but not by heat treatment. They have excellent ductility, weldability and corrosion resistance.

Ferritic These alloys have body-centered atomic structure. They cannot be hardened by heat treatment and only moderately by cold-working. They have good ductility and corrosion resistance, reasonable weldability, and generally higher strengths than the austenitic alloys.

Duplex These alloys contain a mixed crystal structure of austenite and ferrite. They can be hardened by cold-working, and have excellent ductility, excellent corrosion resistance, and generally greater strengths than austenitic alloys. They have good weldability and metal fatigue properties.

Many different numbering systems of stainless steel alloys have been adopted, particularly in Europe. This diversity can create confusion among structural engineers, given that stainless steel is not a frequently used material in structural design. The confusion is aggravated by the fact that certain alloys are included in some systems but not in others.

The systems used in Australia and the United States are the American Iron and Steel Institute system (AISI 1974b) and the Unified Numbering System (ASTM 2002). The EN-10088 Standard (EN-10088 1995) was prepared by the European Committee for Standardisation to unify the European numbering systems and specify material compositions and properties of the alloys included. Table **1** provides a correlation between the three numbering systems for the alloys most commonly used in structural applications.

The AISI system is used predominantly in this chapter. According to this system, the alloys of the 200 and 300 series are austenitic, while the alloys of the 400 series are ferritic. However, the AISI system does not include duplex alloys. The suffixes L, N and M are used to indicate extra low carbon content, added nitrogen and added molybdenum respectively. While these modifications to the basic alloys may have significant influences on the weldability, hardness and corrosion resistance, they affect only slightly the mechanical properties. For instance, the mechanical properties of AISI 304L and 316L are approximately the same as those of AISI 304 and 316 respectively, although in this case slightly lower because of the reduced carbon content.

5.2.3 Selection

Despite the large number of stainless steel alloys available, the number of alloys likely to be considered for structural applications is relatively limited. Stainless steel with very high alloy content is unlikely to be selected because of its high costs, and many alloys are deemed unsuitable because of lack of ductility or weldability. Table 1 lists the alloys included in the main structural design guidelines (ANSI/ASCE-8 1991; Burgan 1993; Eurocode3-1.4 1996).

Table 1. Stainless steel alloys included in ANSI/ASCE-8 (1991), Eurocode3, 1.4, (1996) and AS/NZS4673 (2001)

Alloy			Group*	Specification#
AISI	UNS	EN10088		
201	S20100		A	ASCE-8
205	S20500		A	ASCE-8
301	S30100		A	ASCE-8
301LN	S31053		A	EC3-1.4
302	S30200		A	ASCE-8
304	S30401	1.4301	A	ASCE-8, EC3-1.4, AS/NZS
304L	S30403	1.4306	A	ASCE-8, EC3-1.4, AS/NZS
304LN	S30453	1.4311	A	ASCE-8, EC3-1.4
306		1.4362	A	ASCE-8, EC3-1.4
316	S31600	1.4401	A	ASCE-8, EC3-1.4, AS/NZS
316L	S31603	1.4404/1.4435	A	ASCE-8, EC3-1.4, AS/NZS
316LN	S31653	1.4311	A	ASCE-8, EC3-1.4
316Ti	S31635	1.4571	A	ASCE-8, EC3-1.4
317	S31700		A	ASCE-8
317L	S31703		A	ASCE-8
317LMN		1.4439	A	EC3-1.4
321	S32100	1.4541	A	EC3-1.4
405	S40500		F	ASCE-8
409	S40900		F	ASCE-8, AS/NZS
429	S42900		F	ASCE-8
430	S43000		F	ASCE-8, AS/NZS
439			F	ASCE-8
		14003	F	EC3-1.4, AS/NZS
2205	S31803	1.4462	D	EC3-1.4, AS/NZS

* A – Austenitic, F – Ferritic, D – Duplex
# ASCE-8 – ANSI/ASCE-8 (1991),
EC3-1.4 – Eurocode3, Part 1.4, (1996)
AS/NZS – AS/NZS4673 (2001)

Choosing an appropriate alloy for a particular application requires a thorough understanding of stainless steel. This is particularly true for construction in highly corrosive environments (eg chemical industries) for which expert advice should generally be sought. Nevertheless, Table 2 gives an overview of likely structural applications of stainless steel alloys in corrosive environments. Further information can be found in (Lula 1965; NiDI 1990; AISI 1991a).

The 304L and 316L stainless steel alloys are widely used for structural applications. The 304L alloy is relatively inexpensive, and is resistant to rust and pitting under normal atmospheric conditions. The 316L alloy is suited for structural applications in lightly corrosive marine environments. The alloys are easily fabricated by machining and welding.

Generally, stainless steel alloys are not as readily available as carbon steel grades. Hence, an important part of the selection process is to ensure availability of the alloy and to make certain that it can be delivered within the construction timeframe.

Table 2. Selection of stainless steel based on corrosion resistance (NiDI 1990)

Alloy	Location															
	Rural				Urban				Industrial				Seaside			
	I	L	M	H	I	L	M	H	I	L	M	H	I	L	M	H
302, 304, 304L	√	√	√	√	√	√	√	∇	√	∇	∇	∇	√	∇	∇	∇
316, 316L	√	√	√	√	√	√	√	√	√	√	√	∇	√	√	∇	∇
430	√	∇	∇	∇	√	×	×	×	∇	×	×	×	∇	×	×	×
S31803	√	√	√	√	√	√	√	√	√	√	√	√	√	√	√	√

I – Indoor

L – Least corrosive conditions (low temperature, low humidity)

M – Medium, typical of each location

H – Highly corrosive conditions (high temperature, high humidity, air pollution)

√ – Suitable

∇ – Unsuitable; however usable if a smooth surface finish material is used and washed frequently

× – Unsuitable

5.3 Material properties

5.3.1 General

From a structural viewpoint, the two main properties that distinguish stainless steels from carbon steel are the stress-strain characteristics and the ability to maintain stiffness and strength at elevated temperatures. In contrast to carbon steel, for which the stress-strain curve may be modelled as bi-linear for most structural applications, the stress-strain curve of stainless steel is generally highly nonlinear and without a distinct yield point. This is demonstrated in Figure 1, which compares typical stress-strain curves of Grade 250 and 350 carbons steel with stress-strain curves for annealed, 1/2- and full-hard 302 and 304 stainless steel. As a result of the nonlinearity, stainless steel loses stiffness at low stress levels. This affects the design rules for members that rely on stiffness to transfer loads, notably compression members and unbraced flexural members.

In the absence of a yield stress, it is common practice to define an equivalent yield stress for stainless steel by using a proof stress, usually the 0.2% proof stress. (By definition, the plastic - or permanent - strain of the 0.2% proof stress is 0.2%). The proof stresses of many stainless steel alloys are higher than the yield stresses of Grades 250 and 350 carbon steel, and so stainless steel may offer superior strength-to-weight ratios.

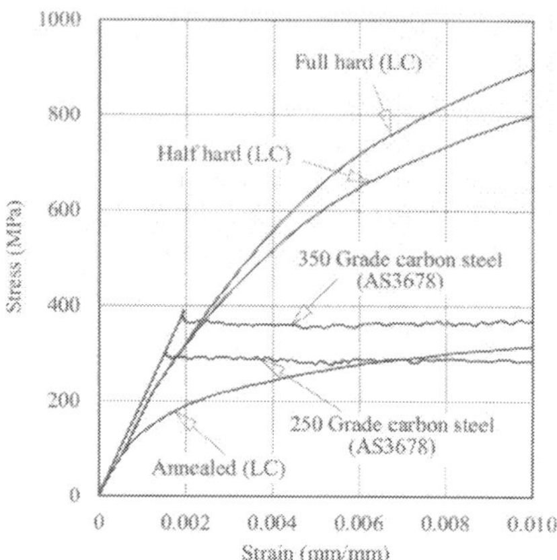

Figure 1. Stress-strain curves of Grade 250 and 350 carbon steel, and annealed, 1/2- and full-hard AISI 302 and 304 stainless steel. LC indicates "longitudinal compression".

5.3.2 Mechanical properties

It is possible to enhance the strength of austenitic stainless steel by cold-working to a much greater extent than for carbon steel. Figures 2a, 2b and 2c show the increases in 0.2% proof stress and ultimate tensile strength produced by cold-reducing AISI 304, 304L and 316 stainless steel. The ductility decreases accordingly, as also shown in Fig 2.

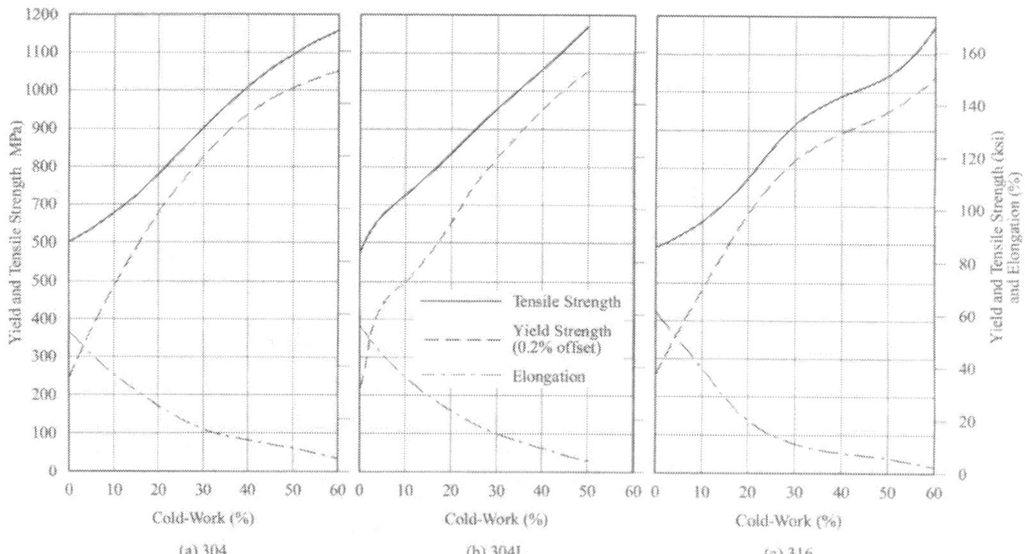

Figure 2. Effect of cold-work on AISI 304, 304L and 316 stainless steel alloys

The pronounced cold-working ability is often utilised by cold-reducing stainless steel plate and coil to various tempers before fabrication. The tempers are referred to as ¼- and ½- and full-hard, according to requirements of minimum proof stress and tensile strength, as specified in material standards, eg (ASTM-A176 1999; ASTM-A276 2000; ASTM-A666 2000; ASTM-A240 2002). (Note that the temper is not directly related to the percentage cold-reduction so that, for instance, ½ -hard does not imply a 50% reduction in thickness). Figures 3a, 3b and 3c show stress-strain curves for annealed, ½- and full-hard AISI 302 and 304 stainless steel alloys respectively.

Cold-working is also utilised in structural design of stainless steel tubes. The design guidelines proposed in (Rasmussen and Hancock 1993a; 1993b) for stainless steel structural hollow sections were based on the properties of the finished cold-formed product. The cold-working of the steel during forming produced compressive and bending strengths of about double the magnitude of those based on the annealed properties. Such increases are unrivaled by carbon steel tubes for which cold-forming enhances the strength by 10-30%. However, given the considerable number of manufacturers of stainless steel tubes and the fact that most of the tubes on the market are annealed after cold-forming, it is imperative that the mechanical properties be certified by the manufacturer if increases in 0.2% proof stress and tensile strength above the annealed values are to be incorporated in design.

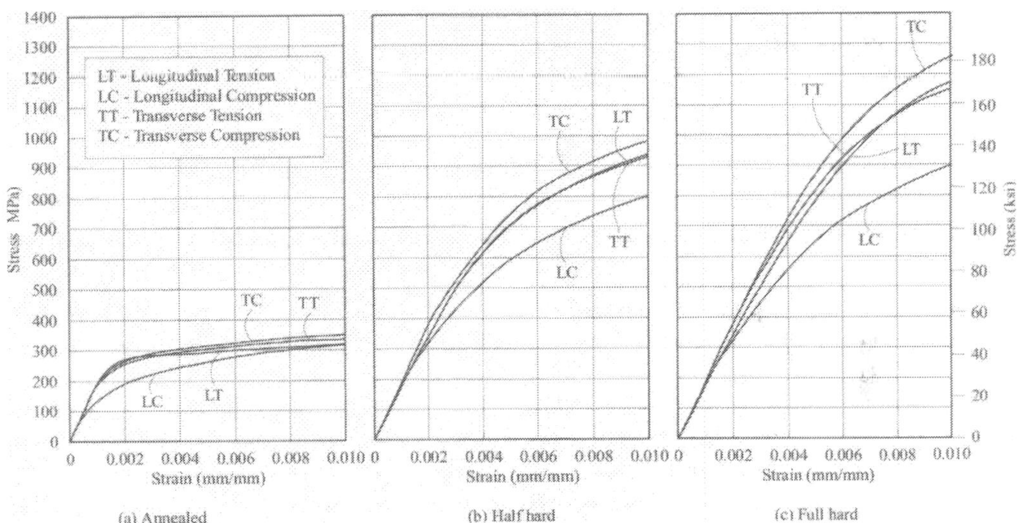

Figure 3. Stress-strain curves for annealed, 1/2- and full-hard 302 and 304 stainless steel alloys.

Stainless steel is anisotropic, (i.e. the mechanical properties are different in different directions), and behaves differently in tension and compression. The anisotropy depends on the degree of cold-work, as shown in Figure 3, and on the fabrication process.

Figure 4a shows stress-strain curves for annealed AISI 304 stainless steel pertaining to the (longitudinal, L) direction of rolling and the (transverse, T) direction perpendicular to this. The coupons were tested in both tension (T) and compression (C). The four stress-strain curves are different, thus demonstrating the anisotropy and the difference between compressive and tensile properties. From a structural viewpoint it is important to notice that the lowest stress-strain curve is that for longitudinal compression (LC), since this is the curve of primary concern in designing compression members and flexural members. Similar stress-strain curves are shown in Figure 4b for the flat part of a cold-formed square tube of AISI 304L stainless steel. In this case, the lowest stress-strain curve is that for transverse tension (TT) rather than longitudinal compression (LC).

The initial modulus of elasticity (E_0) of stainless steel alloys is slightly lower than that of carbon steel, depending on the alloy and on the fabrication process. It is also slightly different in the longitudinal and transverse directions. As a general rule, the initial modulus may be assumed to be E_0=195 GPa which compares with approximately 205 GPa for carbon steel. In design, accurate values of the initial modulus should be used, such as those specified in (ANSI/ASCE-8 1991; AS/NZS4673 2001).

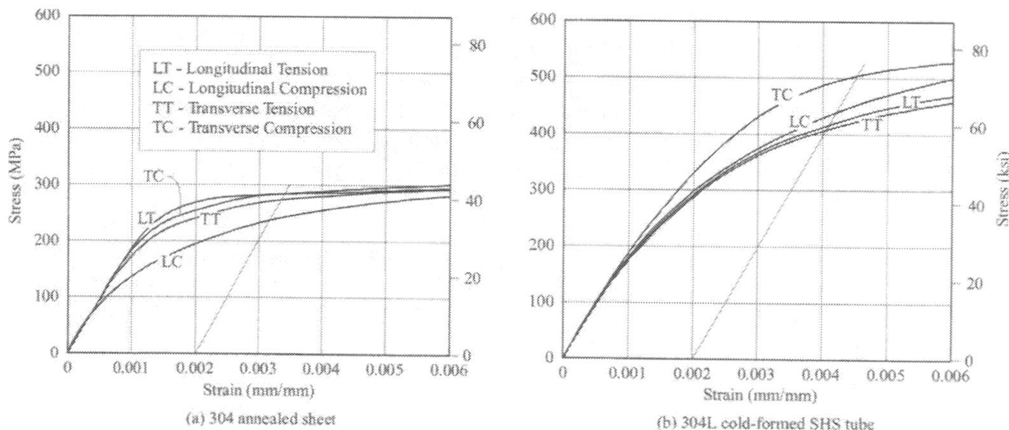

Figure 4. Stress-strain curves for annealed 304 stainless steel sheet and cold-formed AISI 304L stainless steel square tube.

The behaviour of stainless steel at elevated temperatures is superior to that of carbon steel. This is demonstrated in Figures 5a and 5b which show graphs of yield stress (or 0.2% proof stress) and tensile strength as functions of temperature. The graphs are shown for AISI 304, 304L, 316 and 316L stainless steels and for Grade 350 carbon steel. The fact that stainless steel maintains substantial strength at elevated temperatures is important in design against fire and in design of vessels containing hot gasses or liquids.

Detailed information on the chemical composition and mechanical properties of most of the austenitic and ferritic stainless steel alloys that are likely to be used in structural applications can be obtained from (AISI 1974b). The properties at elevated temperatures are covered comprehensively in (Simmons and Echo 1965).

The coefficients of expansion of austenitic stainless steel alloys are generally larger than those of carbon steel, as shown in Table 3. At the same time, the thermal conductivity is lower, as also shown in the table. While the larger coefficient of expansion is important in determining thermally induced stresses and deformations, the combination of larger coefficient of expansion and lower thermal conductivity has the effect of inducing greater welding distortions than those experienced in fabricating carbon steel structural members.

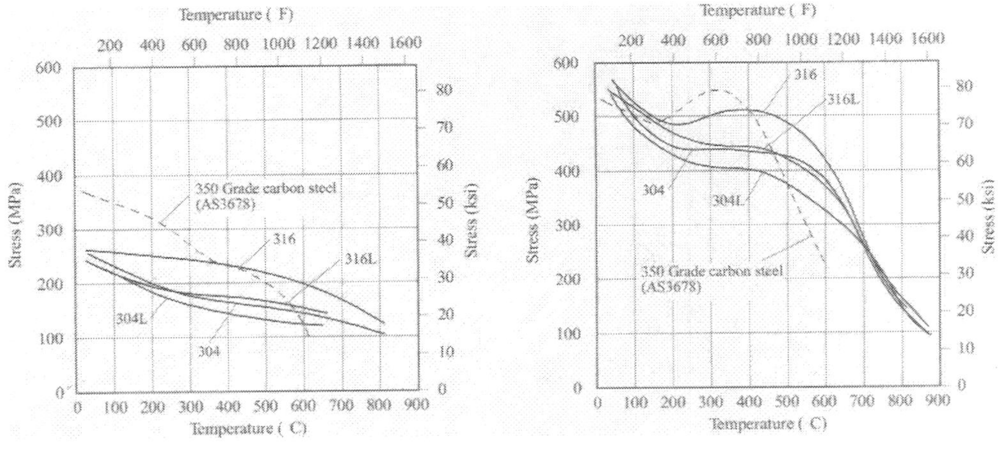

(a) 0.2% Proof stress (b) Tensile strength

Figure 5. Mechanical properties of stainless steel and carbon steel at elevated temperatures (Simmons and Echo 1965)

Table 3. Room temperature values of density, coefficient of expansion, and thermal conductivity of stainless steel alloys and carbon steel (Peckner and Bernstein 1977)

Alloy	Density (kg/m^3)	Lin. Coef. Of expansion (10^{-6})	Thermal conductivity (W/m°C)
201	7700	18.4	16
301	7700	18.0	15
304	8000	18.2	14
316	8000	17.5	14
409	7700	11.7	25
430	7700	11.2	21
S31803	7800	13.7	19
Carbon steel	7850	11.7	58

5.3.3 Analytical expression for the stress-strain curve and its moduli

It is common practice to use the Ramberg-Osgood expression (Ramberg and Osgood 1943),

$$\varepsilon = \frac{\sigma}{E_0} + 0.002\left(\frac{\sigma}{\sigma_{0.2}}\right)^n \tag{1}$$

for modelling the stress strain curve of stainless steel alloys. It involves the initial elastic modulus (E_0), the 0.2% proof stress ($\sigma_{0.2}$) and the parameter (n) which defines the sharpness of the knee of the stress-strain curve, as shown in Figure **6**.

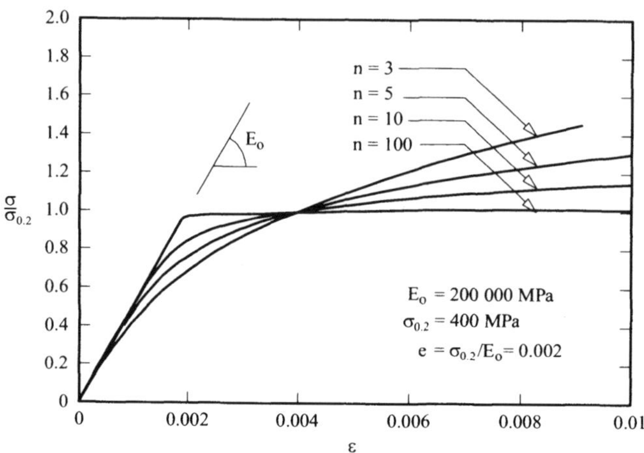

Figure 6. Ramberg-Osgood stress-strain curves

In the limit where $n \rightarrow \infty$, the Ramberg-Osgood expression produces a bi-linear curve (such as that of carbon steel).

The slope of the stress-strain curve, or the tangent modulus (E_t), is frequently used in design calculations as is the secant modulus (E_s), defined as the slope of the line connecting the origo with the current stress point, as shown in Figure 7. Analytical expressions for these moduli are readily obtained from eqn. (1),

$$E_t = \frac{E_0}{1 + 0.002nE_0 / \sigma_{0.2} \left(\frac{\sigma}{\sigma_{0.2}} \right)^{n-1}} \tag{2}$$

$$E_s = \frac{E_0}{1 + 0.002E_0 / \sigma_{0.2} \left(\frac{\sigma}{\sigma_{0.2}} \right)^{n-1}} \tag{3}$$

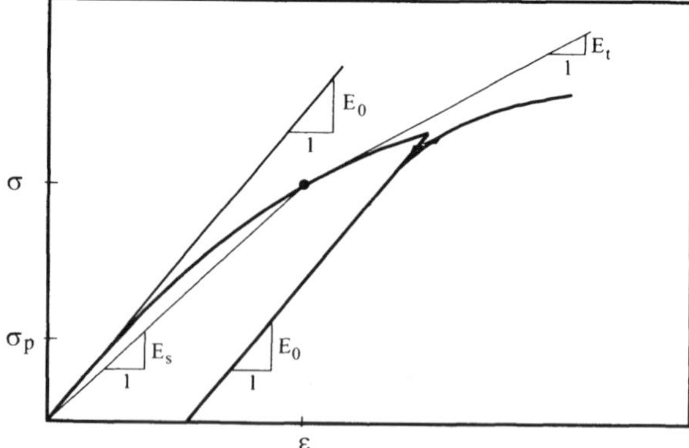

Figure 7. Initial (E_0), tangent (E_t), and secant (E_s) moduli

5.3.4 Mechanical properties for structural design

The American (ANSI/ASCE-8 1991), Australian (AS/NZS4673 2001) and South African (SABS-0162-4 1997) specifications for cold-formed stainless steel structures include tables of mechanical properties for the stainless steel alloys most commonly used in structural practice. The properties include the initial Young's modulus (E_0), the equivalent yield stress (F_y), defined as the 0.2 % proof stress, the ultimate tensile strength (F_u), the proportionality stress (F_p), defined as the 0.01 % proof stress, and the Ramberg-Osgood n-parameter.

The mechanical properties specified in the Australian standard for cold-formed stainless steel structures (AS/NZS4673 2001) are shown in Table **4**. The values are for alloys in their annealed state. The table shows the mechanical properties for compression and tension for loading in the (longitudinal) direction of rolling (see Figure **8**) and the transverse direction, as well as shear. In addition to these properties, the American ANSI/ASCE Specification for the Design of Cold-formed Stainless Steel Structural Members (ANSI/ASCE-8 1991) also includes mechanical properties for 1/16, 1/4 and 1/2 hard tempers of 201, 301, 304 and 316 austenitic alloys.

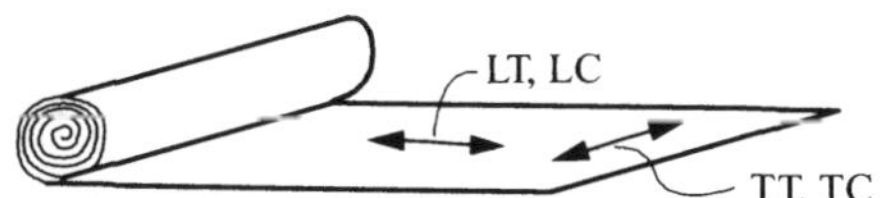

Figure 8. Longitudinal and transverse directions of rolling

Table 4. Mechanical properties as included in Appendix B of AS/NZS4673 (2001)

		304 316	**304L 316L**	**409**	**1.4003 (3Cr12)**	**430**	**S31803 (2205)**
Initial modulus (E_0)	(GPa)	195	195	185	195	185	200
0.2% proof stress (F_y)	(MPa)	205	205	205	250	275	430
Ult. tensile strength (F_u)	(MPa)	520	485	380	435	450	590
Proportionality stress (F_p)	(MPa)	140	140	155	180	195	245
n-parameter		7.5	7.5	11	9	8.5	5.5

a) Longitudinal tension

		304 316	**304L 316L**	**409**	**1.4003 (3Cr12)**	**430**	**S31803 (2205)**
Initial modulus (E_0)	(GPa)	195	195	185	210	185	195
0.2% proof stress (F_y)	(MPa)	195	195	205	260	275	435
Proportionality stress (F_p)	(MPa)	90	90	150	170	170	245
n-parameter		4	4	9.5	7.5	6.5	5

b) Longitudinal compression

		304 316	304L 316L	409	1.4003 (3Cr12)	430	S31803 (2205)
Initial modulus (E_0)	(GPa)	195	195	200	220	200	205
0.2% proof stress (F_y)	(MPa)	205	205	240	280	310	450
Ult. tensile strength (F_u)	(MPa)	520	485	380	460	450	620
Proportionality stress (F_p)	(MPa)	118	118	200	215	250	245
n-parameter		5.5	5.5	16	11.5	14	5

c) Transverse tension

		304 316	304L 316L	409	1.4003 (3Cr12)	430	S31803 (2205)
Initial modulus (E_0)	(GPa)	195	195	200	230	200	205
0.2% proof stress (F_y)	(MPa)	205	205	240	285	310	445
Proportionality stress (F_p)	(MPa)	135	135	200	220	255	265
n-parameter		7	7	16	11.5	15	5.5

d) Transverse compression

		304 316	304L 316L	409	1.4003 (3Cr12)	430	S31803 (2205)
Initial modulus (G_0)	(GPa)	75	75	75	75	75	75
0.2% proof stress (F_{yv})	(MPa)	115	115	130	155	165	255
n-parameter		6	6	13	10	11	5.5

e) Shear

5.3.5 Corrosion

Some guidance to selecting the alloy on the basis of corrosion resistance is given in Table 2. Further information may be obtained from (Lula 1965; CMC 1966; AISI 1974b; Sedriks 1979). However, for specialised applications in highly corrosive environments, expert advice should be sought.

The types of corrosion which in some instances can be encountered in stainless steel structures are summarised in (Eurocode3-1.4 1996; AS/NZS4673 2001) as: pitting, crevice corrosion, bimetallic corrosion, stress corrosion cracking, general corrosion, and intergranular attack (or sensitisation). The same references define these types of corrosion and suggest ways of guarding against them.

5.4 Effect of Material Nonlinearity on Strength and Deformation Calculations

5.4.1 General

In Section 5.3, we saw that stainless steel alloys have nonlinear stress-strain curves with low proportionality stress. At stresses above the proportionality stress, the material yields and loses stiffness gradually. The stiffness is measured as the tangent of the stress-strain curve, as shown in Figure 7, which varies with the stress level. The tangent modulus plays an important role in stability calculations, since the buckling resistance depends on the current stiffness among other factors.

5.4.2 Strength of columns

Engesser (1889) published one of the earliest studies of the effect of yielding on the buckling resistance of columns. He proposed that the first term of the buckling equation for a column,

$$E_0 I \frac{d^4 w}{dx^4} + P \frac{d^2 w}{dx^2} = 0 \tag{4}$$

be changed to reflect the loss of stiffness. The term encapsulates the bending moment that develops during overall buckling and as such is related to the current stiffness of the material. Engesser originally proposed that the bending stress resulting from overall buckling could be assumed to be proportional to E_t. In this case, the buckling equation changes to,

$$E_t I \frac{d^4 w}{dx^4} + P \frac{d^2 w}{dx^2} = 0 \tag{5}$$

and the buckling load is obtained as,

$$P_{E_t} = \frac{\pi^2 E_t I}{L_e^2} \tag{6}$$

where L_e is the effective length which depends on the end support conditions.

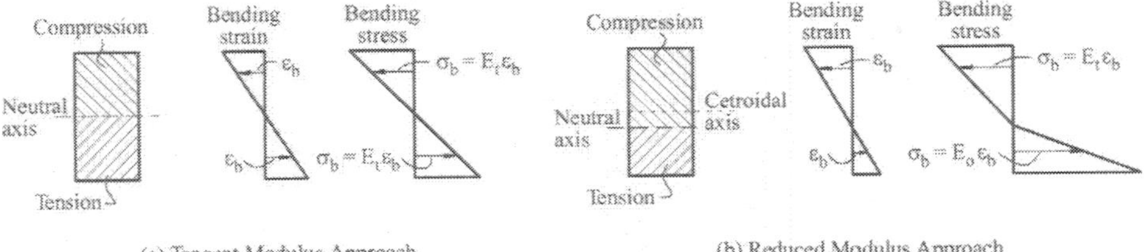

Figure 9. Tangent and reduced moduli approaches

Engesser's first proposal was critisised because it did not recognise that during overall buckling, tension may be induced on the concave side of the columns which therefore would elastically unload, and the initial modulus (E_0) would pertain to this part of the cross-section. In response, Engesser (1895), modified eqn. (6) to account for elastic unloading in the part of the cross-section in tension, as shown in Figure 9. This required a new position of the neutral axis which was determined so that the axial force remained unchanged. The bending moment depended upon the initial elastic modulus (E_0) for the parts of the cross-section in tension and the tangent modulus (E_t) for the parts of the section in compression, as shown in Figure **9**. The buckling load could then be determined as,

$$P_{E_r} = \frac{\pi^2 EI_r}{L_e^2} \tag{7}$$

where the reduced modulus (E_r) was a function of the elastic modulus (E_0) and the tangent modulus (E_t),

$$E_r = E_t \frac{I_c}{I} + E_0 \frac{I_t}{I} \tag{8}$$

In eqn. (8), I_c and I_t are the second moments of area of the parts of the cross-section in compression and tension with respect to the neutral axis, respectively, and I is the second moment of area of the full cross-section with respect to the centroidal principal axis.

However, when compared to test results on concentrically loaded columns, it was found that the tangent modulus approach (eqn. (6)) provided better agreement than the reduced modulus approach. Shanley (1947) used a simple strut model to explain this result by pointing out that:

1. The column cannot remain straight at loads above the tangent modulus buckling load (P_{E_t}), since if it did the tangent stiffness of all point in the cross-section would be E_t and the column would be in a state of unstable equilibrium. The column must therefore start to buckle at P_{E_t}.

2. The axial force cannot be assumed to remain constant during overall buckling but may increase, as shown in Figure 10.

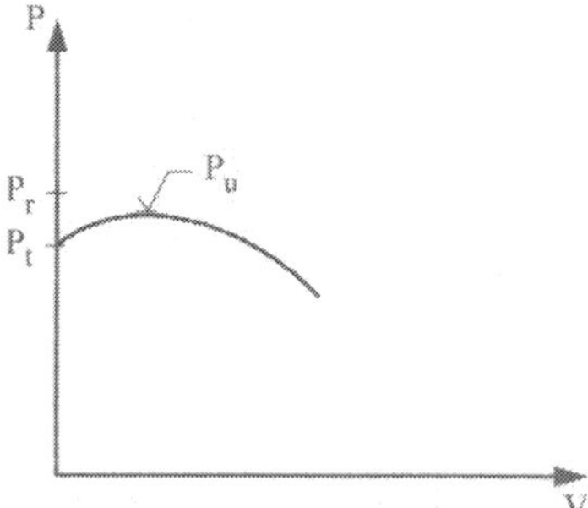

Figure 10. Load-deflection curve

However, the inelastic post-buckling reserve is generally small for columns and so the tangent modulus buckling load (P_{E_t}) is a conservative and yet reasonably accurate measure of the inelastic buckling strength. The tangent modulus is used in the American, South African and Australian standards to determine the strength of compression and unbraced flexural members.

5.4.3 Strength of plates

Stowell (1948) developed a theory for calculating the inelastic buckling load of plates. Based on Shanley's work, he assumed that in the inelastic range, an increase in load proceeds simultaneously so that strain reversal does not occur. For a plate uniformly compressed in the x-direction (see Figure **11**), the following buckling equation was derived,

$$D^{*}\left[\left(1-\tfrac{3}{4}\kappa\right)\frac{\partial^4 w}{\partial x^4}+2\frac{\partial^4 w}{\partial x^2\partial y^2}+\frac{\partial^4 w}{\partial y^4}\right]+t\sigma_x\frac{\partial^2 w}{\partial x^2}=0 \tag{9}$$

where t is the plate thickness and,

$$D^{*}=\frac{E_s t^3}{9} \tag{10}$$

$$\kappa=1-\frac{E_t}{E_s} \tag{11}$$

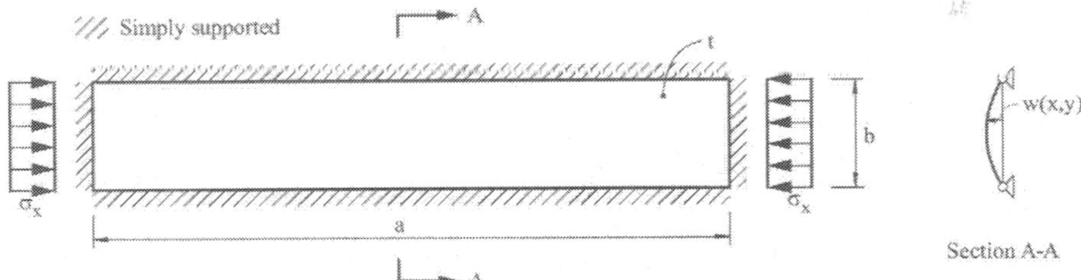

Figure 11. Buckling of rectangular plate under uniform compression

In equations (9-11), E_s is the secant modulus and σ_x is the applied compressive stress. Equation (9) is the equivalent of St Venant's equation for the elastic buckling of plates.

$$D\left[\frac{\partial^4 w}{\partial x^4}+2\frac{\partial^4 w}{\partial x^2\partial y^2}+\frac{\partial^4 w}{\partial y^4}\right]+t\sigma_x\frac{\partial^2 w}{\partial x^2}=0 \tag{12}$$

where

$$D=\frac{E_0 t^3}{12(1-\nu^2)} \tag{13}$$

The equivalence between eqns (9) and (12) follows from the fact that for plastic buckling, Poisson's ratio (ν) is taken as 0.5, and for elastic buckling $E_s = E_t = E_0$. It can be concluded from eqns (9-11) that,

1. The plate buckling stress depends on two moduli (E_s, E_t), and

2. The buckling stress is not simply proportional to D^* because of the factor (1-3/4κ) on the $\partial^4 w/\partial x^4$-term of eqn. (9).

The solution of eqn. (9) is involved and generally not suited for design calculations. Bleich (1952) proposed the simplified version of the equation,

$$D\left[\tau^2\frac{\partial^4 w}{\partial x^4}+2\tau\frac{\partial^4 w}{\partial x^2\partial y^2}+\frac{\partial^4 w}{\partial y^4}\right]+t\sigma_x\frac{\partial^2 w}{\partial x^2}=0 \tag{14}$$

in which τ is a plasticity reduction factor that he chose as,

$$\tau=\sqrt{\frac{E_t}{E_0}} \tag{15}$$

Bleich solved eqn. (14) for several support conditions and obtained the following expression for the minimum buckling stress,

$$\sigma_{cr}=\frac{k\pi^2E_0\tau}{12(1-\nu^2)}\left(\frac{t}{b}\right)^2 \tag{16}$$

where k is the plate buckling coefficient which is independent of τ and takes the same values as for elastic buckling.

When the plasticity reduction factor (τ) is taken as $\sqrt{E_t/E_0}$ (eqn. (15)), the minimum buckling stress can be written as,

$$\sigma_{cr}=\frac{k\pi^2\sqrt{E_0E_t}}{12(1-\nu^2)}\left(\frac{t}{b}\right)^2 \tag{17}$$

Evidently, this equation simplifies to the expression for the elastic buckling stress when assuming E_t=E_0.

By comparing eqns (6,17), it can be seen that the inelastic buckling of plates is less affected by gradual yielding than columns. Furthermore, while the buckling of columns can be assumed to represent the ultimate load for all practical purposes, slender plates are post-buckling stable and can support loads in excess of the inelastic buckling stress. It follows from eqn. (17) that when the buckling stress reduces below the proportionality stress, the tangent modulus approaches the initial elastic modulus ($E_t \rightarrow E_0$), and so the buckling stress becomes the elastic buckling stress. Since the elastic buckling stress is a conservative estimate of plate strength, equation (17) becomes increasingly conservative for predicting plate strength as the slenderness increases.

5.4.4 Deflection calculations

The material softening associated with early yielding increases the deformations when the stress exceeds the proportionality stress. Consequently, in design, there is a greater need for checking that the deflections occurring during normal service do not exceed acceptable limits.

Furthermore, if the section is prone to local buckling, the reduction in flexural stiffness of the locally buckled section must be used in the deflection calculations.

Johnson and Winter (1966) first proposed that the deflection of a beam can be determined using,

$$v = k_v \frac{PL^3}{E_{s_a} I_e} \tag{18}$$

where I_e is the second moment of area of the effective cross-section and E_{s_a} is the average of the secant moduli (E_{s_t}, E_{s_c}) calculated at the extreme fibres in tension and compression at the cross-section of maximum moment,

$$E_{s_a} = \frac{E_{s_t} + E_{s_c}}{2} \tag{19}$$

The secant moduli for tension (E_{s_t}) and compression (E_{s_c}) can be determined using eqn. (3) in conjunction with the Ramberg-Osgood parameters for tension and compression respectively. In eqn. (18), the constant k_v depends upon the loading and support conditions and is defined such that eqn. (18) reproduces the linear-elastic expression for the deflection when E_{s_a} is replaced by E_0. (For instance, the mid-span deflection of a linear-elastic simply supported beam subjected to two point loads acting at a quarter from each end is v=11/192 $PL^3/(E_0 I)$ and so k_v =11/192 for this case).

Since the moment generally varies along the length of a beam, the stress and hence secant modulus vary along the beam as well. (The secant modulus also varies through the depth of the cross-section but its values at the extreme fibres are most important for deflection calculations). Johnson and Winter (1966) proposed that the secant moduli (E_{s_t}, E_{s_c}) be determined at the point of maximum moment. However, when applied to SHS and CHS beams, Rasmussen and Hancock (1993b) observed that the combination of determining the average secant modulus (E_{s_a}) at the section of maximum moment using the stresses at the extreme fibres produced excessive estimates of deflection. They proposed that the secant modulus be determined at the stress,

$$\sigma = k_\sigma \frac{M_{\max}}{S_f} \tag{20}$$

where $M_{\max}$ is the maximum moment in the span, S_f is the elastic modulus and k_σ is a factor less than unity which was obtained from calibration against finite element deflection calculations (Rasmussen and Hancock 1992; 1993b). The values of k_σ=2/3 and k_σ=3/4 were obtained for single span SHS and CHS beams respectively, whereas k_σ=1/2 was obtained for SHS and CHS beams continuous over two spans.

Figures **12**a and **12**b show comparisons of deflections predicted using the modified method proposed by Rasmussen and Hancock (1993b) with tests on SHS and CHS beams respectively. Good agreement can be observed over the loading range shown. The horizontal lines labeled $M_{0.2}/1.85$ are the loads producing maximum moments equal to $M_{0.2}/1.85$ where $M_{0.2}=S_p\sigma_{0.2}$ is the plastic moment based on the 0.2% proof stress and the factor 1.85 is the safety factor for allowable stress design specified in Appendix E of the ANSI/ASCE-8 Specification. These loads are representative of those likely to develop during normal service. It follows that for the beams tested, the beams behave nearly linear-elastically during service loads.

More recent studies on the deflection of stainless steel beams (Mirambell and Real 2000; Chryssantholopoulos and Low 2001) have verified the approach suggested by Rasmussen and Hancock (1993b), and more general methods have been proposed (Chryssantholopoulos and Low 2001).

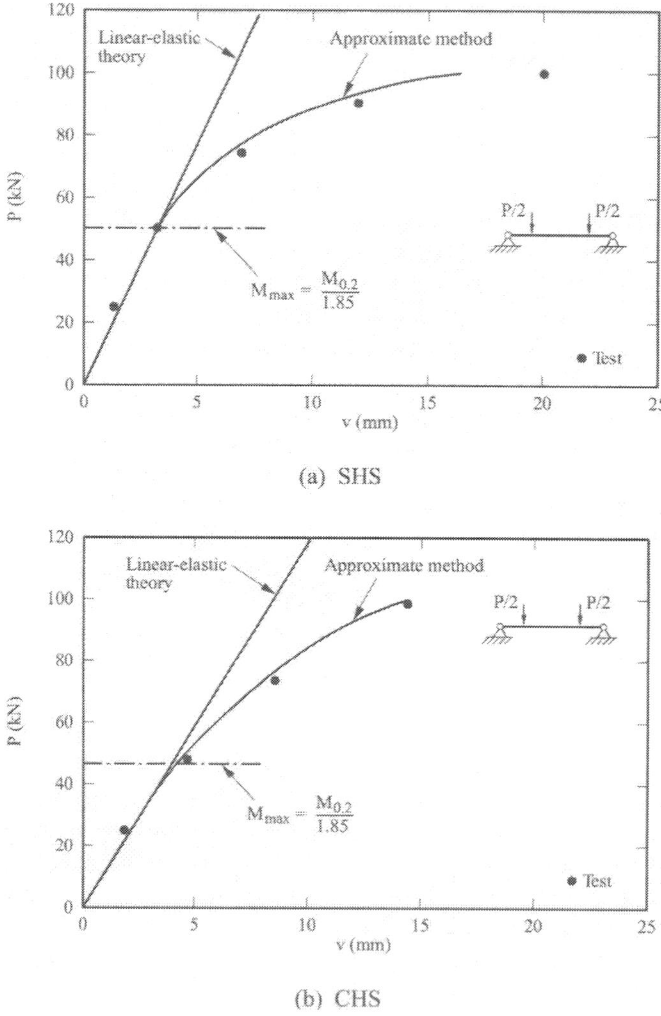

Figure 12. Comparison of experimental and predicted load-deflection curves for SHS and CHS beams (Rasmussen and Hancock 1993b)

5.5 Development of specifications for stainless steel structures

The first American specification (AISI 1968b) for stainless steel structural members was published by the American Iron and Steel Institute in 1968. This applied specifically to annealed and strain-flattened light-gauge steel, cold-formed into structural shapes. A revision of this specification (AISI 1974a), which included 1/4- and 1/2-hard tempers, was published in 1974. The 1968 and 1974 editions were based mainly on the 1968 edition of the AISI specification for cold-formed *carbon* steel structural members (AISI 1968a).

A major revision of the 1968 specification for cold-formed carbon steel structural members was published in 1986 (AISI 1986), prompting a revision of the 1974 cold-formed stainless steel specification. This (Lin et al. 1988a) was prepared at the University of Missouri-Rolla but not published as a specification. Concurrently with the revision of the 1974 stainless steel specification, the 1986 cold-formed carbon steel specification was being converted from allowable stress design format to limit state design format. It was then decided to also convert the stainless steel cold-formed specification to limit state format before publication. Details of the limit state calibration are contained in (Lin et al. 1988b). The same document was published as a specification (ANSI/ASCE-8 1991) by the American Society of Civil Engineers (ASCE) in 1991. The ANSI/ASCE-8 Specification is closely aligned with the AISI limit state specification for cold-formed carbon steel structures (AISI 1997).

In Britain, the Steel Construction Institute prepared a guide (Burgan 1993) on the design of stainless steel structures. The guide is essentially an addendum to the British steel structures standard (BSI 2001), containing design rules that are specific to stainless steel. (It also contains useful guidance on selection and corrosion of stainless steel structures). More recently, the British Steel Institute published a similar guide incorporating research undertaken primarily in Europe during the 1990s (Baddoo and Burgan 2001).

In a similar approach to the British guidelines, the draft Part 1.4 of Eurocode3 (Eurocode3-1.4 1996) for stainless steel structures contains design rules to be used in conjunction with those of Parts 1.1 (Eurocode3-1.1 1992) and 1.3 (Eurocode3-1.3 1996) for hot-rolled and cold-formed carbon steel structures. Thus the British and European approaches are *not* to have separate standards for stainless steel structures, but a limited set of rules that replace those for carbon steel structures for only the types of members that are affected by differences in mechanical properties of stainless and carbon steel.

Standards Australia recently published a joint Australian-New Zealand standard (AS/NZS4673 2001) for the design of stainless steel cold-formed structures. The standard is based on the ANSI/ASCE-8 Specification but includes mechanical properties for a wider range of alloys, as well as additional provisions for tubular members and welded connections. The notation and layout of the standard are the same as those for the Australian/New Zealand standard for cold-formed carbon steel structures (AS/NZS4600 1996), which is based on the limit state edition of the AISI Specification (AISI 1991b).

5.6 Design of Compression Members

5.6.1 General

The strength of compression members (or columns) is usually determined as a product of an overall buckling stress and an area, which may be an effective area if the section is prone to local buckling. In this approach, the local buckling stress is implicitly determined in calculating the effective area. The local and overall buckling modes are the most common modes encountered in design. However, if the column is likely to buckle in other modes in its ultimate limit state including the distortional mode, such buckling modes must be accounted for as well.

In calculating the local and overall buckling stresses, it is necessary to consider the loss of stiffness associated with early yielding. This can be done by using tangent and/or secant moduli in the buckling strength equations as discussed in Section 5.4. In this case, the design procure is implicit (the strength calculations are iterative). Alternatively, strength curves may be used which are lower than those for sections not subject to early yielding. In this case, the strength calculations are direct (the design procedure is explicit). In the following sections, the implicit and explicit formulations are explained in detail.

5.6.2 Implicit Approach

5.6.2.1 Overall Buckling Strength

As discussed in Section 5.4, the ultimate limit stress (σ_u) can be taken as the inelastic buckling stress $\sigma_{E_t} = P_{E_t} / A$ determined using the tangent modulus approach,

$$\sigma_u = \frac{\pi^2 E_t}{(L_e / r)^2} \tag{21}$$

where r is the radius of gyration,

$$r = \sqrt{\frac{I}{A}} \tag{22}$$

This procedure is valid for doubly and point-symmetric sections which fail by flexural buckling. It generally leads to slightly conservative design strengths compared to tests on *concentrically* loaded columns. This is demonstrated in Figure **13** which compares tests on cold-formed circular sections (Rasmussen and Hancock 1993a) with design strengths obtained using eqn. (21).

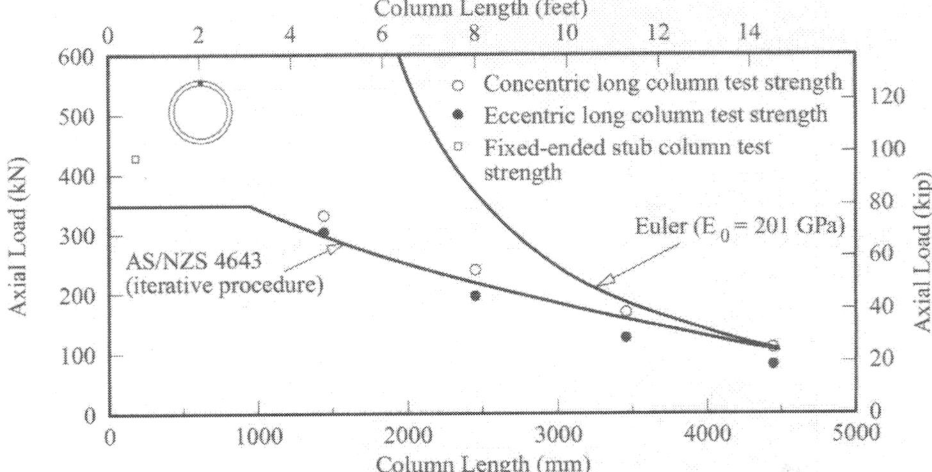

Figure 13. Comparison of tangent modulus design strength with tests (Rasmussen and Hancock 1993a) on CHS columns

The tangent modulus strength curves for the most commonly used structural alloys are shown in Figure **14**. The curves were obtained by using the Ramberg-Osgood parameters for longitudinal compression shown in Table **4** in conjunction with eqn. (2) for calculating the tangent modulus. The strength is nondimensionalised with respect to the squash load, $\chi=\sigma_u/\sigma_{0.2}$, and the column length is given as nondimensionalised slenderness $\lambda=\sqrt{\sigma_{0.2}/\sigma_{E0}}$, where σ_{E0} is the Euler buckling stress based on the initial elastic modulus (E_0). Figure **14** also shows the Eurocode3, Part 1.4, strength curve for cold-formed sections.

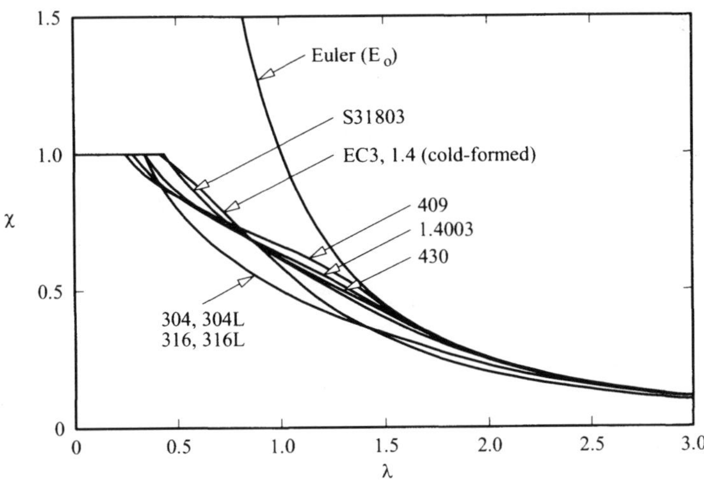

Figure 14. Nondimensional strength curves obtained using tangent modulus theory, and Eurocode3-1.4 strength curve for cold-formed sections

The tangent modulus procedure can also be used for singly symmetric sections subject to flexural-torsional buckling, such as the channel section shown in Figure **15**. In this case, the ultimate strength is determined as,

$$\sigma_u = \frac{1}{2\beta}\left(\sigma_{ex} + \sigma_t - \sqrt{(\sigma_{ex} + \sigma_t)^2 - 4\beta\sigma_{ex}\sigma_t}\right) \tag{23}$$

where σ_{ex} is the major axis flexural buckling strength obtained using eqns (21-22) with $r=r_x$ in eqn. (22), σ_t is the torsional buckling strength and β is a cross-sectional parameter which depends on the polar radius of gyration (r_0) and distance between the centroid and the shear centre (x_0),

$$\beta = 1 - \left(\frac{x_0}{r_0}\right)^2 \tag{24}$$

$$r_0 = \sqrt{\frac{I_x + I_y}{A}} \tag{25}$$

The torsional buckling strength (σ_t) is given by

$$\sigma_t = \frac{1}{Ar_0^2}\left(G_0 J + \frac{\pi^2 E_0 I_\omega}{l_{et}^2}\right)\frac{E_t}{E_0} \tag{26}$$

where G_0 is the initial shear modulus, J is the torsion constant, I_ω is the warping constant and l_{et} is the effective torsional length. The term E_t/E_0 in eqn. (26) accounts for the effect of yielding. It is assumed in this equation that the uniform torsion stiffness (GJ) is reduced by the same factor as the warping torsion stiffness (EI_ω).

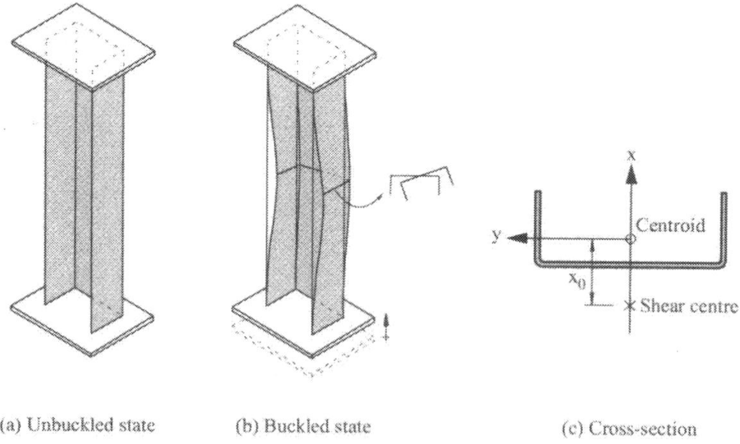

Figure 15. Channel section column undergoing flexural-torsional buckling

The tangent modulus approach is iterative because the tangent modulus is to be determined at the stress (σ_u) which is unknown. An initial guess of (σ_u) is therefore made and the tangent modulus (E_t) is calculated at this stress, typically using eqn. (2) for given values of the Ramberg-Osgood parameters (E_0, $\sigma_{0.2}$, n). A new buckling stress is calculated using eqn. (21) or (23) and this process is repeated until convergence is obtained.

The tangent modulus approach assumes the column buckles from a straight position and so ignores the effect of initial overall crookedness, (also referred to as overall geometric imperfections). To account for the effect of geometric imperfections, the American and Australian design specifications apply a relatively small resistance factor (ϕ) to the buckling strength. According to these specifications, the column strength is calculated as,

$$P_u = \phi A_e \sigma_u \tag{27}$$

where A_e is the effective area and the resistance factor is taken as,

$$\phi = 0.85 \tag{28}$$

5.6.2.2 Effective Area Calculation (Local Buckling)

The effective area (A_e) is the sum of the effective part of each component plate and the corner areas (A_c) which can be assumed fully effective,

$$A_e - \sum b_e t + \sum A_c \tag{29}$$

In eqn. (29), b_e is the effective width which is less than or equal to the full width (b). The effective width depends on the buckling stress of the component plates.

According to Bleich's simplified theory, the inelastic buckling stress (σ_{cr}) is reduced by the same factor ($\sqrt{E_t/E_0}$) for plates supported along both longitudinal edges (stiffened elements) and plates supported along one longitudinal edge (unstiffened elements), see eqns (15,16). However, the buckling modes for stiffened and unstiffened plates are different in that the former involves significant flexure transversely while the latter does not, and experiments have shown that different plasticity factors apply to the two types of elements (Gerard 1946). Van den Berg (2000) studied the effect of local buckling of stainless steel plate elements and proposed that the plasticity factors shown in Table **5** should be used for calculating the plate buckling stress. The buckling stresses thus obtained are compared with tests in Fig. **16** and seen to be in reasonable agreement.

Table 5. Plasticity reduction factors (τ) for stiffened and unstiffened elements (Berg 2000)

	Plasticity reduction factor (τ)
Stiffened Elements	$\sqrt{\frac{E_t}{E_0}}$
Unstiffened Elements	$\frac{E_s}{E_0}$

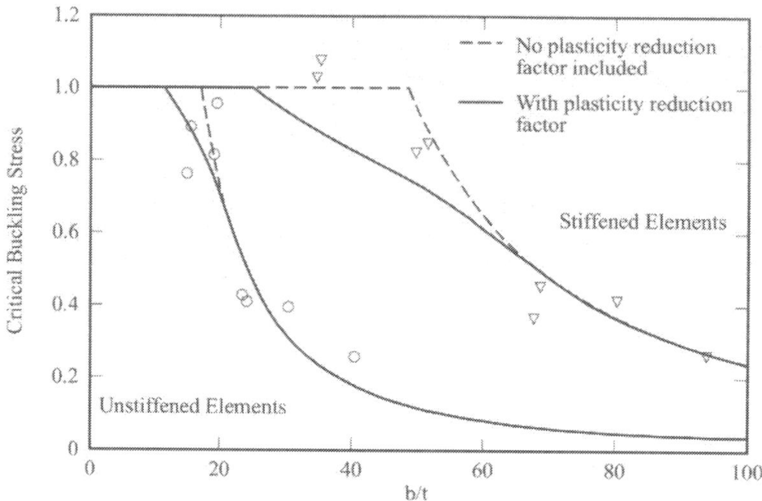

Figure 16. Buckling stress vs slenderness

Van den Berg further proposed that the effective width of plates in uniform compression should be determined using the Winter expression,

$$\frac{b_e}{b} = \begin{cases} 1 & \text{for } \lambda \le 0.677 \\ \dfrac{1 - 0.22/\lambda}{\lambda} & \text{for } \lambda > 0.677 \end{cases} \tag{30}$$

where

$$\lambda = \sqrt{\frac{\sigma}{\sigma_{cr}}} \tag{31}$$

$$\sigma_{cr} = \frac{k\pi^2 E_0 \tau}{12(1-\nu^2)}\left(\frac{t}{b}\right)^2 \tag{32}$$

In eqns (31-32), σ is the applied stress, k is the plate buckling coefficient equal to 4 and 0.425 for stiffened and unstiffened plates respectively, ν is the Poisson's ratio taken as 0.3, and τ is the plasticity reduction factor given in Table **5** which is to be calculated at the stress σ.

It should be noticed that when the term $0.22/\lambda$ is ignored in eqn. (30), the effective width is proportional to $1/\lambda$ and hence $\sqrt{\tau}$. This implies that the plate strength is less affected by yielding than the inelastic buckling stress (σ_{cr}) which is proportional to τ.

5.6.3 Explicit Approach

5.6.3.1 General

From Section 5.4, we know that the local and overall buckling strengths are affected by early yielding and that the reduction in strength is related to the stress-strain curve. We also know that different alloys have different stress-strain curves and so, different strength curves apply to different alloys. Since the stress-strain curve can be represented closely by the Ramberg-Osgood expression (eqn. (1)), it is possible to express the strength curve directly in terms of the Ramberg-Osgood parameters.

Rasmussen and Rondal (1997) developed the relationship for columns failing by flexural buckling. More recently, Rasmussen et al. (2002) have developed the relationship for plates.

Table 6: (n, e)-combinations in Rasmussen and Rondal (1997)

n	3	5	10	20	40	100		
e	0.001	0.0015	0.002	0.0025	0.003	0.004	0.006	0.008
(n,e)	(3, 0.001), (3, 0.0015),... (3, 0.008),.. (5, 0.001), (5,0.0015),..., (100, 0.008)							

5.6.3.2 Overall Buckling Strength

Rasmussen and Rondal (1997) used the finite element program developed by Clarke (1994) to obtain strength curves for the combinations of the n-parameter and the nondimensional proof stress ($e=\sigma_{0.2}/E_0$) shown in Table **6**. The column strength (P_u) was normalised wrt the "yield" load ($P_{0.2}$),

$$P_{0.2} = A\sigma_{0.2} \tag{33}$$

and expressed in terms of the column slenderness,

$$\lambda = \sqrt{\frac{P_{0.2}}{P_{E_0}}} \tag{34}$$

where P_{E_0} is the *elastic* Euler buckling stress,

$$P_{E_0} = \frac{\pi^2 E_0 I}{l_e^2} \tag{35}$$

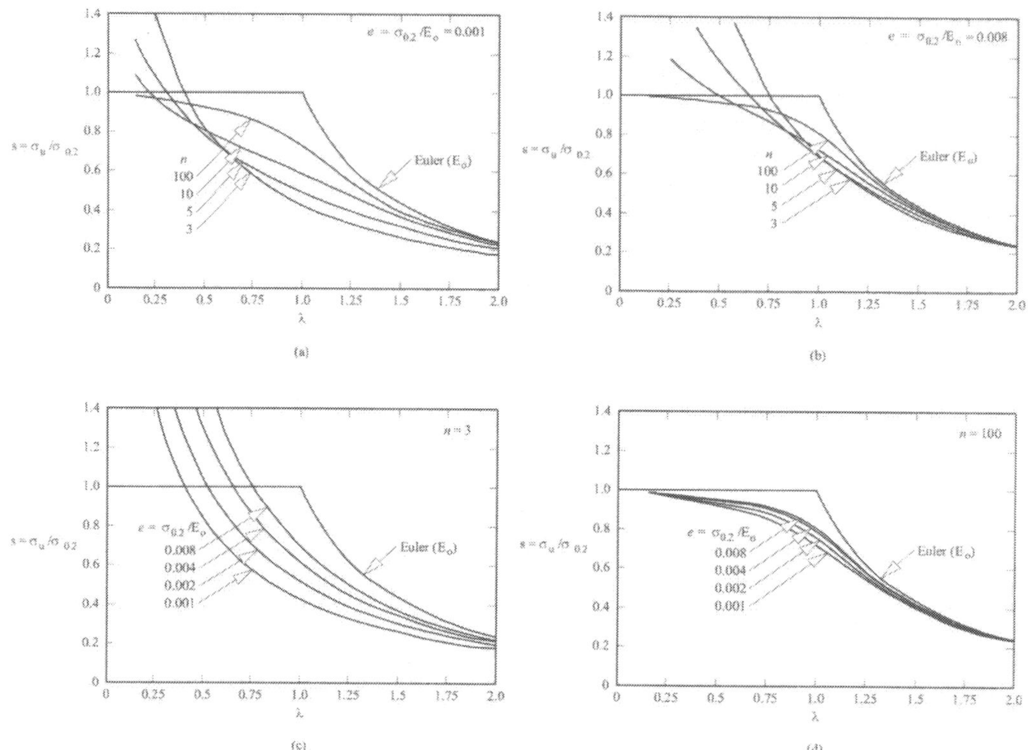

Figure 17. Finite element column strength curves

Figures **17**a and **17**b show strength curves for a range of n-values for e=0.001 and e=0.008 respectively. Similarly, Figures **17**c and **17**d show strength curves for a range of e-values for n=3 and n=100 respectively, where n=100 represents carbon steels with a bi-linear stress-strain curve.

The following observations can be made from Figures **17**a-**17**d:

- There is a significant difference in strength for different values of n for low values of e but not for large values of e, (cmp. Figs 17a and 17b).
- There is significant difference in strength for different values of e for low values of n but not for large values of n, (cmp. Figs 17c and 17d). Consequently, the proof stress has much greater effect on the strength curves for small values of n than for large values of n.
- For $\lambda \to 0$, the strength exceeds the yield load for small values of n, while for larger values of n the strength approaches the yield load. This result is a consequence of the stress-strain curves detailed in Fig. 6 which show that the stress may significantly exceed the proof stress at large strain for small values of n.
- For low values of n, the strength curves have single curvature while for large value of n, they have double curvature, (cmp. Figs 17c and 17d).

To obtain an analytic expression for the strength strain curve, Rasmussen and Rondal (1997) adopted a Perry curve in the form

$$\chi = \frac{1}{\varphi + \sqrt{\varphi^2 - \lambda^2}} \le 1 \tag{36}$$

where χ is the nondimensional column strength, and φ is defined as,

$$\varphi = \frac{1}{2}\left(1 + \eta + \lambda^2\right) \tag{37}$$

In eqn (37), η is the imperfection parameter which in standards for carbon steels, such as Eurocode3, Part 1.1 (Eurocode3-1.1 1992), is assumed to be linear in λ,

$$\eta = \alpha(\lambda - \lambda_0) \tag{38}$$

To account for the influence of early yielding, Rasmussen and Rondal (1997) used the nonlinear relationship,

$$\eta = \alpha\left((\lambda - \lambda_1)^{\beta} - \lambda_0\right) \tag{39}$$

and expressed α, β, λ_0 and λ_1 in terms of the parameters n and e, as follows:

$$\alpha(n,e) = \frac{1.5}{\left(e^{0.6} + 0.03\right)\left(n^{\left(\frac{0.0048}{e^{0.55}}\right)+1.4} + 13\right)} + \frac{0.002}{e^{0.6}} \tag{40}$$

$$\beta(n,e) = \frac{0.36\exp(-n)}{e^{0.45} + 0.007} + \tanh\left(\frac{n}{180} + \frac{6\times 10^{-6}}{e^{1.4}} + 0.04\right) \tag{41}$$

$$\lambda_0(n,e) = 0.82\left(\frac{e}{e + 0.0004} - 0.01n\right) \ge 0.2 \tag{42}$$

$$\lambda_1(n,e) = 0.8\frac{e}{e+0.0018}\left(1-\left[\left(\frac{n-5.5}{n+\dfrac{6e-0.0054}{e+0.0015}}\right)^2\right]^{0.6}\right) \qquad (43)$$

Figures **18**a-**18**d show comparisons of the finite element strength curves with the strength curves obtained using eqns (36-43). The maximum discrepancy at any slenderness value (λ) for any of the values of (*n*,*e*) shown in Table **6** was 5.6% which implies close agreement between the curves.

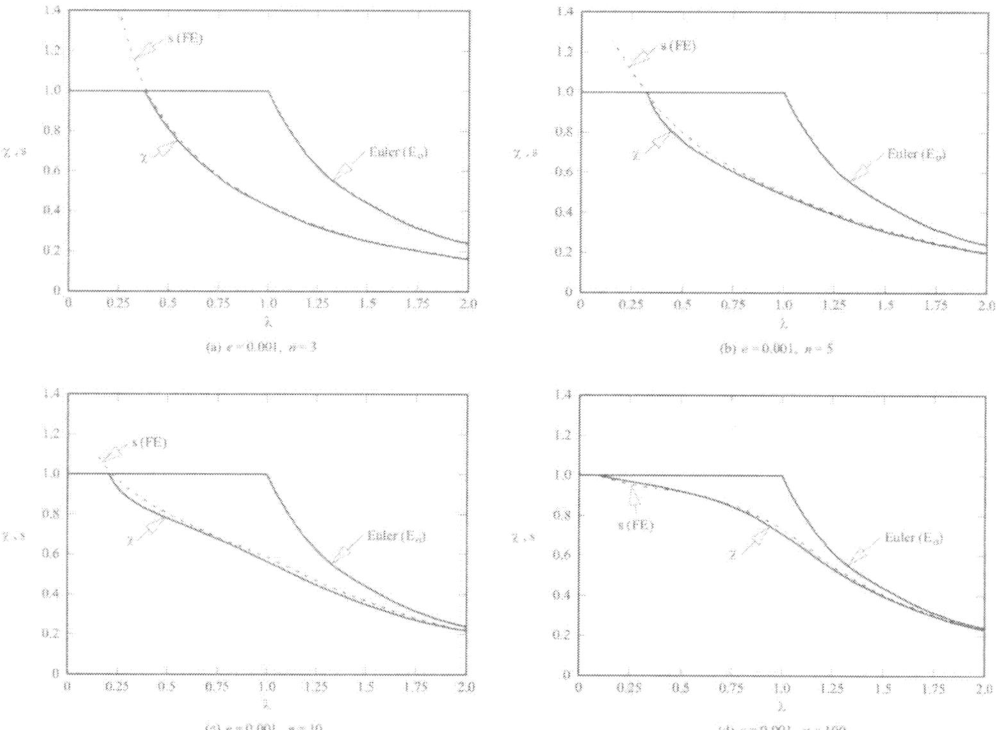

Figure 18. Comparison of finite element strength curves with curve fits

In applying eqns (36-43) to the design of stainless steel columns, the Ramberg-Osgood parameters for longitudinal compression given in Table **4** were used to compute values of α, β, λ_0 and λ_1 as shown in Table 7. These values combined with eqns (36,37,39) allow a direct calculation of the nondimensional column strength (χ). The resulting strength curves are shown in Fig. 19. The strength curve for cold-formed sections given in Part 1.4 of Eurocode3 for stainless steel members is also shown in Figure **19**. The explicit approach has been incorporated in the Australian Standard (AS/NZS4673 2001).

Table 7: Values of α, β, λ_0 and λ_1

	304 316	**304L 316L**	**409**	**1.4003 (3Cr12)**	**430**	**S31803 (2205)**
α	1.59	1.59	0.77	0.94	1.04	1.16
β	0.28	0.28	0.19	0.15	0.14	0.13
λ_0	0.55	0.55	0.51	0.56	0.59	0.65
λ_1	0.20	0.20	0.19	0.27	0.33	0.42

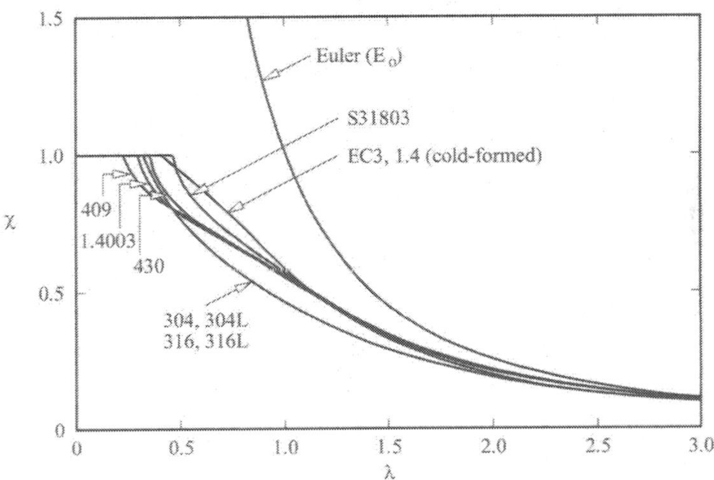

Figure 19. Column strength curves obtained using explicit column curve formulation, and Eurocode3-1.4 strength curve for cold-formed sections

5.6.3.3 Section Strength

Rasmussen et al. (2002) used the finite element program Abaqus (Hibbitt et al. 1995) to obtain strength curves for square plates under uniform compression. The plates were simply supported along all four edges and had geometric imperfections in the shape of the elastic buckling mode with a magnitude of a tenth of the thickness. The stress-strain curve was defined by the Ramberg-Osgood parameters (E_0, $\sigma_{0.2}$, n) using the extended Ramberg-Osgood expression (Rasmussen 2001).

Finite element analyses were carried out at several plate slenderness values ($\lambda=\sqrt{\sigma_{0.2}/\sigma_{cr}}$) for each of the combinations of n and e shown in Table 8. The slenderness was based on the *elastic* buckling stress (σ_{cr}) calculated using eqn (32) with k=4 and τ=1. The nondimensional strength (s) was obtained as the ratio of the ultimate load (P_u) to the yield load ($P_{0.2}$) defined by eqn. (33),

$$s = \frac{P_u}{P_{0.2}} \tag{44}$$

Table 8. (n, e)-combinations in (Rasmussen et al. 2002)

n	3	5	10	100		
e	0.001	0.0015	0.002	0.0025	0.003	
(n,e)	(3, 0.001), (3, 0.0015),… (3, 0.003),.. (5, 0.001), (5,0.0015),…, (100, 0.003)					

Figures 20a and 20b show strength curves for several values of n for e=0.001 and e=0.003 respectively. Similarly, Figures 20c and 20d show strength curves for several values of e for n=3 and n=100 respectively. It follows from Figs 20a and 20b that the lower the value of n, the lower the nondimensional strength. It is also clear that the influence of n is less at high values of nondimensional yield stress (e) than it is at lower values. Figures 20c and 20d show that the nondimensional strength is dependent on the proof stress for low values of n but hardly so for large values of n. These results are consistent with those obtained for columns.

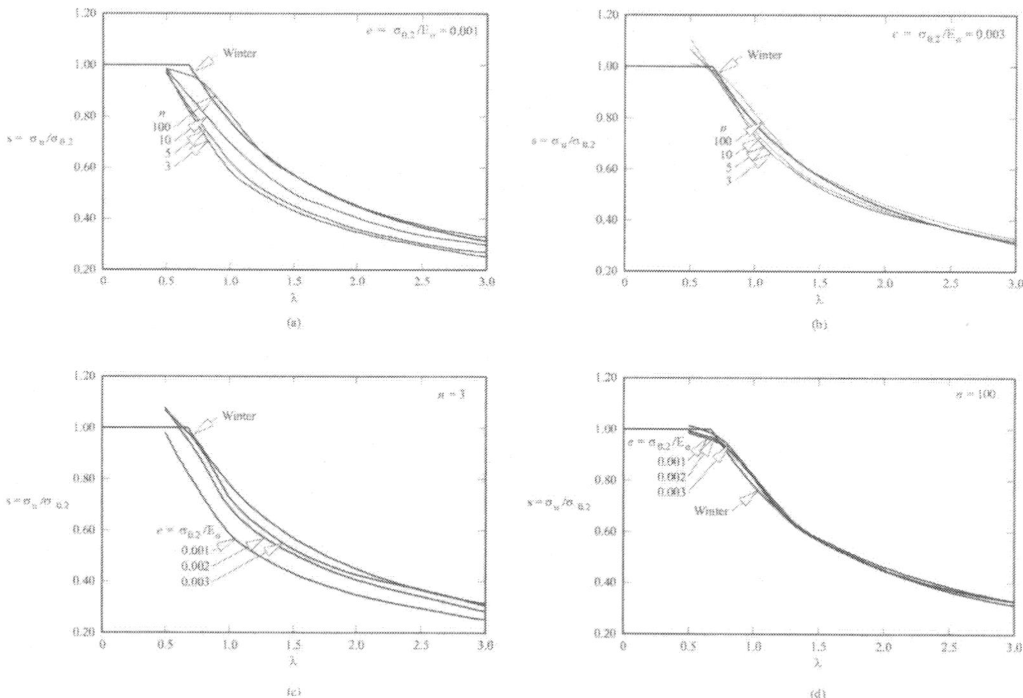

Figure 20. Finite element plate strength curves

The Winter plate strength curve was used as basis in developing analytic approximations to the finite element strength curves. Accordingly, the nondimensional strength (χ) was expressed as,

$$\chi = \alpha / \lambda - \beta / \lambda^2 \leq 1 \quad (45)$$

where the terms α and β are functions of e and n, as follows,

$$\alpha = 0.92 + 0.07\tanh\left(\frac{n-3}{2.1}\right) - \left(0.026\exp[-0.55(n-3)] + 0.019\right)\left(6 - 2000e\right) \text{ for } 3 \le n \le 10 \quad (46)$$

$$\beta = 0.18 + 0.045\tanh\left(\frac{n-3}{2.5}\right) - \left(0.01\exp[-1.6(n-3)] + 0.005\right)\left(6 - 2000e\right) \text{ for } 3 \le n \le 10 \quad (47)$$

Equations (46,47) are valid for $3 \le n \le 10$. Linear interpolation can be used for $10 < n \le 100$ between the values produced by eqns (46,47) for $n=10$ and the values of $\alpha=1$ and $\beta=0.22$ applicable for $n=100$ (representing carbon steel). The $\alpha=1$ and $\beta=0.22$ values are those of the Winter-expression (eqn. (30)) applicable to cold-formed carbon steel.

Figures **21**a-d show comparisons of the finite element strength curves with the strength curves obtained using eqns (45-47).

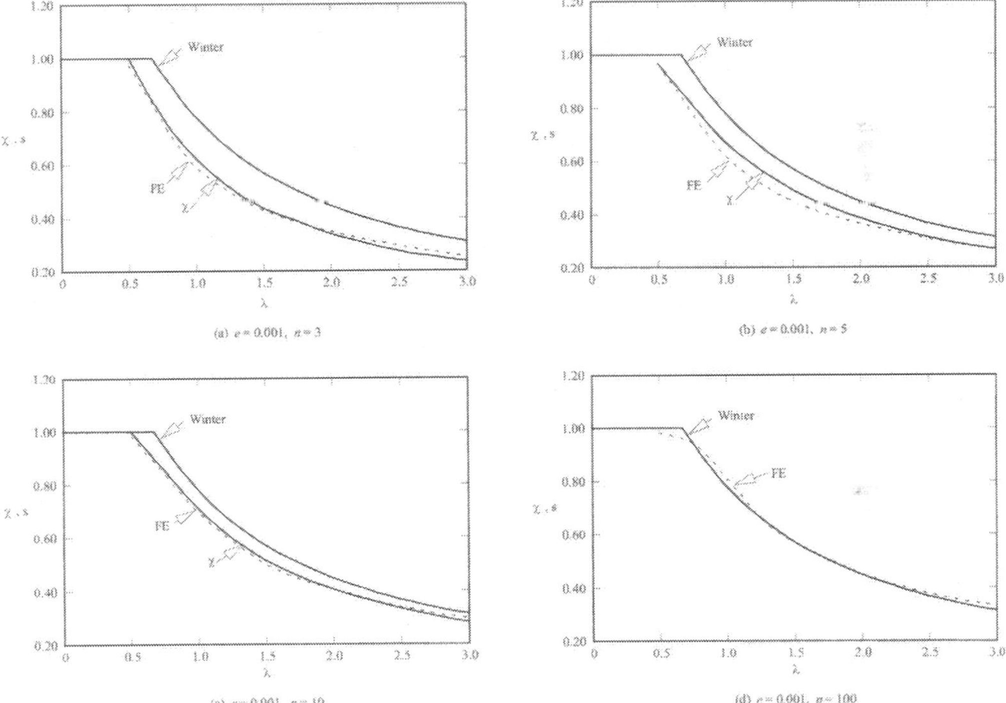

Figure 21. Comparison of finite element strength curves with curve fits

5.6.4 Design specifications

5.6.4.1 General

The main standards for stainless steel structures are the American Specification for the Design of Stainless Steel Structural Members (ANSI/ASCE-8 1991), the joint Australian/New Zealand Standard entitled Cold-formed Stainless Steel Structures (AS/NZS4673 2001) and Eurocode3, Part 1.4 Supplementary Rules for Stainless Steel Structural Members (Eurocode3-1.4 1996).

The American and Australian specifications are targeted at cold-formed structures whereas the European standard refers to Parts 1.1 and 1.3 of Eurocode3 and as such applies to fabricated as well as cold-formed structures.

The Australian/New Zealand Standard AS/NZS4673 is largely based on the American ANSI/ASCE-8 Specification. However, additional design rules are included based on recent research, notably on the flexural buckling of columns and tubular members and connections. The Australian/New Zealand Standard also includes mechanical properties for ferritic EN 1.4003 and duplex UNS31803 alloys.

The American and Australian specifications use the implicit tangent modulus approach to determine the member capacity of columns and unbraced beams. However, for columns failing by flexural buckling, the Australian standard allows the explicit design method detailed in Section 5.6.3 to be used as an alternative method. The European standard uses an explicit approach.

5.6.4.2 Overall Buckling Strength

American and Australian/New Zealand specifications. According to Section 3.4 of the ANSI/ASCE-8 Specification and Section 3.4.1 of AS/NZS4673, the design axial strength (P_d) shall be determined as

$$P_d = \phi_c P_n \tag{48}$$

where

$$\phi_c = 0.85 \tag{49}$$

$$P_n = A_e F_n \tag{50}$$

In equation (50), A_e is the effective area calculated at the stress F_n where F_n is the least of the flexural, torsional and flexural-torsional buckling stresses given by eqns (21), (26) and (23) respectively. It should be noticed that the calculation of F_n is based on the full cross-section rather than the effective cross-section.

In place of the implicit procedure, Section 3.4.2 of AS/NZS4673 allows the buckling stress (F_n) of columns failing by flexure to be calculated using,

$$F_n = \frac{F_y}{\varphi + \sqrt{\varphi^2 - \lambda^2}} \le F_y \tag{51}$$

$$\lambda = \frac{l_e}{r}\sqrt{\frac{F_y}{\pi^2 E_0}} \tag{52}$$

where φ and η are defined by eqns (37) and (39) respectively and the equivalent yield stress (F_y) is the 0.2% proof stress. The calculation is explicit. The corresponding resistance factor is

$$\phi_c = 0.9 \tag{53}$$

European standard. According to Section 5.4.1 of Eurocode3, Part 1.4, the design strength of columns failing by flexure shall be determined as,

$$P_d = \frac{A_e \chi F_y}{\gamma_M} \tag{54}$$

where

$$\gamma_M = 1.1 \tag{55}$$

$$\chi = \frac{1}{\varphi + \sqrt{\varphi^2 - \lambda^2}} \leq 1 \tag{56}$$

$$\varphi = \tfrac{1}{2}\left(1 + \eta + \lambda^2\right) \tag{57}$$

$$\lambda = \frac{l_e}{r}\sqrt{\frac{F_y A_e}{\pi^2 A E_0}} \tag{58}$$

In equation (57), the imperfection parameter (η) is given by the *linear* expression,

$$\eta = \alpha(\lambda - \lambda_0) \tag{59}$$

where for cold-formed sections α=0.49 and λ_0=0.40. The linear expression for η is the same as that used to define the column strength curves in Parts 1.1 and 1.3 of Eurocode3 but the values of α and λ_0 differ to account for the effect of material softening. The European standard uses the same strength curve for all austenitic and duplex alloys. The design of ferritic alloys shall be according to Annex D of Eurocode3, Part 1.4. It should be noticed that according to eqn. (58), the column slenderness (λ) is defined in terms of the effective area.

The European standard does not specify supplementary rules for columns failing by torsional or flexural-torsional buckling. This implies that such members shall be designed according to Parts 1.1 and 1.3 of Eurocode3 for hot-finished and cold-formed sections respectively, and that no reduction shall be applied to such members to account for the effect of material softening.

5.6.4.3 Effective area calculation

American and Australian/New Zealand specifications. According to Sections 2.2 and 2.3 of the ANSI/ASCE-8 Specification and AS/NZS4673, the effective widths (b_e) of stiffened and unstiffened elements shall be calculated as,

$$b_e = \begin{cases} b & \text{for } \lambda \leq 0.673 \\ \rho b & \text{for } \lambda > 0.673 \end{cases} \tag{60}$$

where b is the flat width of the element, ρ is the effective width factor,

$$\rho = (1 - \frac{0.22}{\lambda})/\lambda \tag{61}$$

and λ is the plate slenderness defined as,

$$\lambda = \frac{1.052}{\sqrt{k}}\frac{b}{t}\sqrt{\frac{F}{E_0}} \tag{62}$$

In eqn. (62), F is the applied stress which for compression members is the buckling stress (F_n), and k is the local buckling coefficient which is to be taken as,

$$k = \begin{cases} 4 & \text{for stiffened elements} \\ 0.5 & \text{for unstiffened elements} \end{cases} \tag{63}$$

The effective width calculation uses the Winter formula in a same version as that used in the AISI and AS/NZS4600 specifications for cold-formed carbon steel members. However, the plate buckling coefficient (k) is taken as 0.5 for stainless steel unstiffened elements whereas the AISI and AS/NZS4600 specifications use a value of 0.43.

It should be noticed that according to eqn. (62), which involves the *initial* modulus (E_0), the effective width calculation is based on the elastic buckling stress and so does *not* account for early yielding.

Once the effective widths of the component elements have been obtained, the effective area is calculated using eqn. (29).

European standard. The effective width (b_e) is also calculated using the Winter equation (61) in the European standard but according to Section 5.2.3 of Eurocode3, Part 1.4, the yield stress (rather than the applied stress) is used in calculating the plate slenderness,

$$\lambda = \frac{1.052}{\sqrt{k}} \frac{b}{t} \sqrt{\frac{\gamma_M F_y}{E_0}} \tag{64}$$

The plate buckling coefficient (k) equals 4 and 0.43 for stiffened and unstiffened elements in uniform compression respectively. As in the American and Australian/New Zealand specifications, Eurocode3, Part 1.4, makes no reduction to account for material softening in calculating effective widths.

5.6.4.5 Design Example, Columns

The design strength of the concentrically loaded pin-ended SHS column shown in Figure **22** is calculated in this example according to the American, Australian/New Zealand and European specifications. The cross-section is doubly-symmetric and so overall failure is governed by flexural buckling. The column has been cold-formed from AISI304 austenitic steel and annealed after forming.

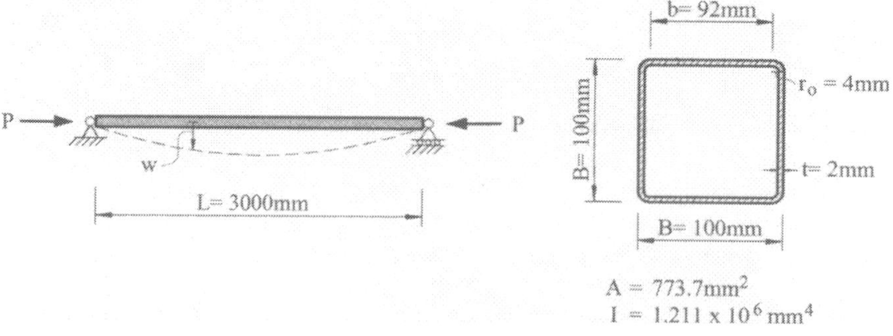

Figure 22. Pin-ended SHS column

American and Australian/New Zealand specifications. The Ramberg-Osgood parameters for annealed AISI304 stainless steel are given in Table B1(B) of AS/NZS4673 as follows,

$$E_0 = 195\,\text{GPa} \quad F_y = 195\,\text{MPa} \quad n = 4$$

These values apply to compression stresses in the longitudinal direction of rolling.

We may assume E_t=E_0/2 =97.5 GPa for the first iteration, and proceed to calculate F_n using eqn. (21),

$$F_n = \frac{\pi^2 E_t}{(L/r)^2} = \frac{\pi^2 \times 975000}{(3000/39.6)^2} = 167.4\,\text{MPa}$$

where $r = \sqrt{I/A}$=39.6 has been used for the radius of gyration.

The tangent modulus at this stress is determined using eqn. (2),

$$E_t = \frac{E_0}{1 + 0.002nE_0 / F_y \left(\frac{F_n}{F_y}\right)^{n-1}}$$

$$= \frac{195}{1 + 0.002 \times 4 \times 195000 / 195 \left(\frac{167.4}{195}\right)^3} - 32.2\,\text{GPa}$$

which is less than the assumed value of E_t=97.5 GPa. We now repeat the calculation with the new value of E_t, and continue the iteration to arrive at

$$E_t = 69.3\,\text{GPa}$$

$$F_n = 118.9\,\text{MPa}$$

The effective area is calculated using eqn. (29). The effective widths of the flat parts are obtained as,

$$b = B - 2r_o = 100 - 2 \times 4 = 92\,\text{mm}$$

$$\lambda = \frac{1.052}{\sqrt{k}} \frac{b}{t} \sqrt{\frac{F_n}{E_0}} = \frac{1.052}{\sqrt{4}} \frac{92}{2} \sqrt{\frac{118.9}{195000}} = 0.598 \le 0.673$$

$$\rho = 1$$

$$b_e = b = 92\,\text{mm}$$

Thus, the section is fully effective at the stress F_n=118.9 MPa and so,

$$A_e = A = 773.7\,\text{mm}^2$$

The column design strength is now obtained using eqns (48-50),

$$P_n = A_e F_n = 773.7 \times 118.9\,\text{N} = 92.0\,\text{kN}$$

$$P_d = \phi_c P_n = 0.85 \times 92.0 = 78.2\ \text{kN}$$

According to the alternative explicit method described in Section 3.4.2 of AS/NZS4673, the column buckling stress can be determined using eqns (51-52), where the imperfection parameter (η) is given by eqn (39) with the following values,

$$\alpha = 1.59 \quad \beta = 0.28 \quad \lambda_0 = 0.55 \quad \lambda_1 = 0.2$$

as obtained from Table 3.4.2 of AS/NZS 4673. Thus,

$$\lambda = \frac{l_e}{r}\sqrt{\frac{F_y}{\pi^2 E_0}} = \frac{3000}{39.6}\sqrt{\frac{195}{\pi^2 \times 195000}} = 0.763$$

$$\eta = \alpha\left((\lambda - \lambda_1)^\beta - \lambda_0\right) = 1.59\left((0.763 - 0.2)^{0.28} - 0.55\right) = 0.479$$

$$\varphi = \tfrac{1}{2}(1 + \eta + \lambda^2) = \tfrac{1}{2}(1 + 0.479 + 0.763^2) = 1.031$$

$$F_n = \frac{F_y}{\varphi + \sqrt{\varphi^2 - \lambda^2}} = \frac{195}{1.031 + \sqrt{1.031^2 - 0.763^2}} = 113.1\ \text{MPa}$$

Since this value of F_n is less than that (F_n =118.9 MPa) obtained from the implicit calculation, the section is again fully effective,

$$A_e = A = 773.7\ \text{mm}^2$$

The column design strength is (now using ϕ_c=0.9),

$$P_n = A_e F_n = 773.7 \times 113.1\ \text{N} = 87.5\ \text{kN}$$

$$P_d = \phi_c P_n = 0.9 \times 87.5\ \text{kN} = 78.8\ \text{kN}$$

European standard. In using Eurocode3, Part 1.4, we start with the effective area calculation,

$$b = B - 3t = 100 - 3 \times 2 = 94\ \text{mm}$$

$$\lambda = \frac{1.052}{\sqrt{k}}\frac{b}{t}\sqrt{\frac{\gamma_M F_y}{E_0}} = \frac{1.052}{\sqrt{4}}\frac{94}{2}\sqrt{\frac{1.1 \times 195}{195000}} = 0.820$$

$$\rho = (1 - \frac{0.22}{\lambda})/\lambda = (1 - \frac{0.22}{0.82})/0.82 = 0.892$$

$$b_e = \rho b = 0.892 \times 94 = 83.8\ \text{mm}$$

$$A_e = A - 4(b - b_e)t = 773.7 - 4 \times (94 - 83.8) \times 2 = 692.5\ \text{mm}^2$$

We now compute the column slenderness,

$$\lambda = \frac{l_e}{r}\sqrt{\frac{F_y A_e}{\pi^2 A E_0}} = \frac{3000}{39.6}\sqrt{\frac{195 \times 692.5}{\pi^2 \times 773.7 \times 195000}} = 0.731$$

$$\eta = \alpha(\lambda - \lambda_0) = 0.49(0.731 - 0.4) = 0.162$$

$$\varphi = \tfrac{1}{2}\left(1 + \eta + \lambda^2\right) = \tfrac{1}{2}\left(1 + 0.162 + 0.731^2\right) = 0.848$$

$$\chi = \frac{1}{\varphi + \sqrt{\varphi^2 - \lambda^2}} = \frac{1}{0.848 + \sqrt{0.848^2 - 0.731^2}} = 0.782$$

$$\eta = \alpha(\lambda - \lambda_0) = 0.49(0.731 - 0.4) = 0.162$$

$$P_d = \chi A_e F_y = 0.782 \times 692.5 \times 195\ \text{N} = 108.1\ \text{kN}$$

$$P_d = \frac{P_n}{\gamma_M} = \frac{108.1}{1.1} = 98.3\ \text{kN}$$

Comparison. The calculations shown in the example above have been repeated for several lengths to obtain column curves for the cross-section shown in Figure 22 as produced by the American, Australian/New Zealand and European specifications. The calculations have been made for the yield stress of annealed AISI304 alloy (F_y=195 MPa), as used in the example above, as well as a yield stress of F_y=400 MPa, which is a typical as-finished value of yield stress of cold-rolled AISI304 hollow sections (Rasmussen and Hancock 1993a; Talja and Salmi 1995). The column strength curves are shown in Figure 23.

At short lengths, the buckling stress is F_n=F_y for all three specifications. The differences in strength result from different ϕ- and γ_M-factors, and a different definition of the flat width (b) in the European standard compared to the American and Australian/New Zealand specifications.

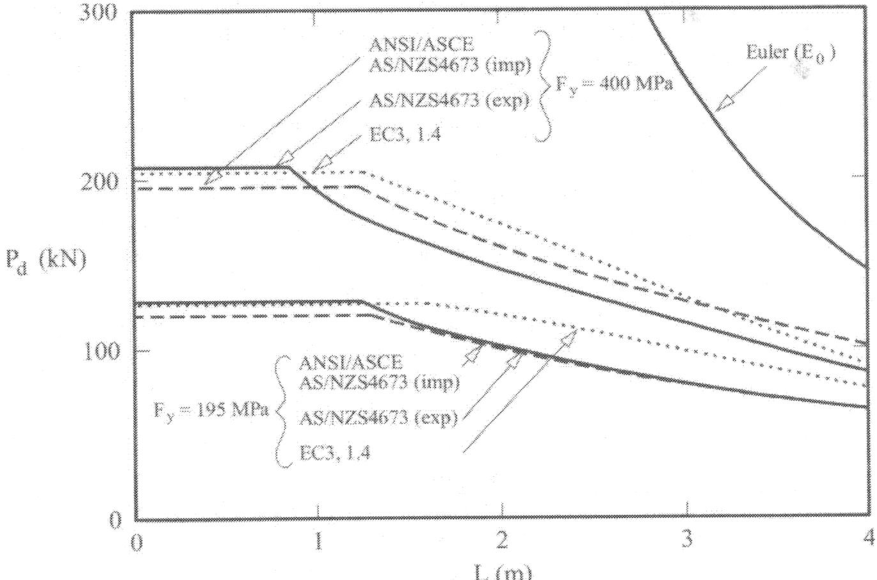

Figure 23. Column curves for F_y=195 MPa and F_y =400 MPa

For F_y=195 MPa, the implicit and explicit formulations of the AS/NZS4673 standard produce almost the same design strengths while the European strength curve is higher because it does not recognise that the strength curve depends on the yield stress (see Fig. 17a) and was calibrated in part on the basis of tests with high values of proof stress, as discussed in detail in Rasmussen and Rondal (2000). For F_y=400 MPa, the strength curve obtained using the implicit formulation of the ANSI/ASCE-8 and AS/NZS4673 specifications is higher than the curves obtained using the explicit formulation and Eurocode3, Part 1.4, at long lengths. This is explained by the fact that the implicit formulation does not account for the effect of overall geometric imperfections, and so approaches the Euler-curve at long lengths. Conversely, the explicit and European column curve formulations were calibrated on the basis of tests and finite element analyses which featured overall geometric imperfections.

5.7 Connections

5.7.1 General

The main methods for joining stainless steel components are bolting, welding, riveting and screwing as for carbon steel. A variety of fasteners, including bolts, welding electrodes and screws, have been developed for general and specific applications, including connections in corrosive environments. The fasteners are made from stainless steel alloys which generally have slightly higher alloying content than the components they join. In the case of welding electrodes, the carbon content is usually low to avoid carbide precipitation, often referred to as sensitization which is the process where carbon combines with chromium to form chromium carbide. Sensitisation occurs when austenitic stainless steel alloys are heated in or cooled through the temperature range 425-900°C. It is undesirable because chromium carbide has reduced corrosion resistance. However, despite these differences, it is no more difficult to bolt or weld stainless steel components than carbon steel components. International specifications are readily available for welding stainless steel, including specifications for stainless steel electrodes and pre-qualified weld preparations (AS/NZS1554.6 1994; AWS-C1.1 2000). Specifications are also readily available for stainless steel bolts (ISO-3506 1997; ASTM-F593 2001).

The main differences between joining stainless and carbon steel are related to detailing and fabrication, rather than structural design. For instance, care needs to be exercised when joining dissimilar metals, such as stainless steel and carbon steel, to avoid galvanic corrosion. A carbon steel screw joining stainless steel sheeting will, for example, quickly corrode if moisture is present to start a galvanic process. Likewise, stainless steel needs to be handled with greater care than carbon steel in fabrication workshops to avoid "contamination", which is when iron particles become embedded in the stainless steel surface and initiate corrosion. The issue of contamination also affects the tools used for fabricating stainless steel products.

Contamination can be avoided provided precautions are taken. For instance, it is good practice to isolate areas where stainless steel products are produced from areas where carbon steel products are produced in fabrication workshops, and to use dedicated stainless steel tools, such as stainless steel wire brushes. Much information is available on the subject of the fabrication of stainless steel, *eg* (NiDI 1992), and reference is made to this material for further information.

The early yielding characterizing stainless steels does not affect the ultimate strength of connections, since connection failure usually occurs at large strains. Furthermore, stainless steel alloys are generally very ductile, particularly the austenitic grades, and premature fracture is unlikely to occur. However, at service loads, larger deformations should be expected which may require a separate calculation.

5.7.2 Structural design

5.7.2.1 Design philosophy

The general design approach for steel connections is to use models of load transfer which satisfy equilibrium and do not exceed the yield stress or the tensile strength. The design models are thus lower bounds and generally produce safe designs. However, the models usually do not consider compatibility and it is therefore implicit that the material has sufficient ductility to develop the necessary strains. This assumption is valid for all common structural stainless steel alloys including austenitic, ferritic and duplex alloys, and the same models as used for carbon steel connections can therefore be applied to stainless steel connections.

The cost of connection design is a function of material and labour costs among other factors. Because of the high material cost of stainless steel, the design of stainless steel connections lends itself less to standardization than carbon steel connections. A more complex detail may be economical if it reduces the material cost through structural efficiency, in spite of higher labour cost.

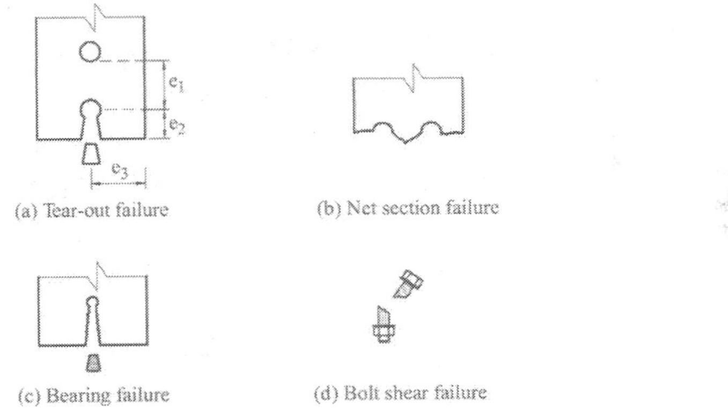

Figure 24. Failure modes in bolted connections

5.7.2.2 Design of bolted connections

American and Australian/New Zealand specifications for cold-formed stainless steel structures.

The design rules of the American and Australian specifications are identical except that the Australian standard includes mechanical properties for bolts to ISO3506 (1997) in addition to bolts to ASTM Standards (ASTM-A276 2000; ASTM-A193 2001; ASTM-F593 2001).

(A) Shear. As in the design of carbon steel, the four basic modes of failure of stainless steel bolted connections subject to shear are those shown in Figs 24a-24d. The load required to produce each of these failure modes is calculated and the ultimate connection strength is taken as the lowest value.

1. Tear-out failure. The design shear force shall be determined as,

$$V_{fd} = \phi V_{fn} \tag{65}$$

where

$$\phi = 0.7 \tag{66}$$

$$V_{fn} = teF_{ut} \tag{67}$$

In equation (67), F_{ut} is the tensile strength of the connected part transverse to the direction of the applied force, t is the thickness of the thinnest connected part, and e is the smaller distance measured in the line of the applied force from the center of the hole to the nearest edge of an adjacent hole or to the end of the connected part, as shown in Fig. **25**.

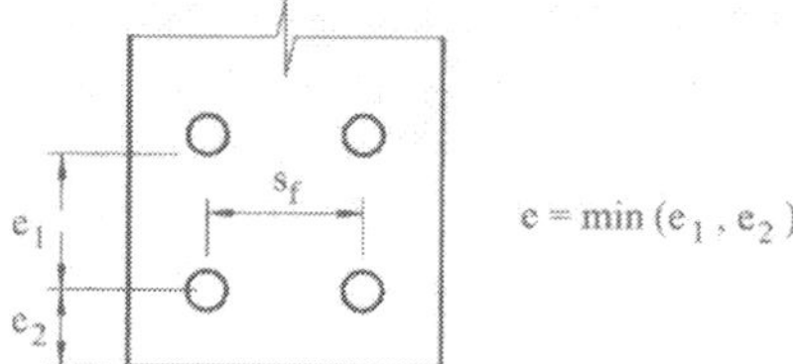

Figure 25. Definition of e and s_f

The distance from the center of the hole to the edge in any direction is required to be larger than $1.5d_f$, where d_f is the diameter of the bolt, and the distance between centers of holes shall not be less than 3 d_f.

Additional requirements apply to oversize and slotted holes.

2. Net section failure. The design force in the connected part shall be determined as,

$$N_{fd} = \phi N_{fn} \tag{68}$$

where

$$\phi = 0.7 \tag{69}$$

$$N_{fn} = A_n F_t \tag{70}$$

In equation (70), A_n is the net area of the connected part in the line of bolts transverse to the line of the applied force and the strength (F_t) is given by,

$$F_t = \begin{cases} (1 - r_f + 2.5 r_f d_f / s_f F_u \leq F_u & \text{for single shear connections} \\ (1 - 0.9 r_f + 3 r_f d_f / s_f F_u \leq F_u & \text{for double shear connections} \end{cases} \tag{71}$$

where s_f is the spacing between the bolts transverse to the line of the force (or in the case of a single bolt, the width of the connected part), see Fig. **25**, and r_f is the force transmitted by the bolt or bolts at the section considered, divided by the tensile force in the member at that section. If r_f is less than 0.2, it may be taken as zero. The procedure for calculating r_f is shown in the example in Section 5.7.4.

As set out under Tension Members in AS/NZS4673 and the ANSI/ASCE-8 Specification, the tensile capacity of the connected member shall be limited to,

$$N_{fyd} = \phi N_{fyn} \tag{72}$$

where

$$\phi = 0.85 \tag{73}$$

$$N_{fyn} = A_n F_y \tag{74}$$

Equation (74) checks that general yielding does not occur. It should be noticed that the net area (A_n) is used in this calculation rather than the gross area.

3. Bearing failure. The design bearing force shall be determined as,

$$V_{bd} = \phi V_{bn} \tag{75}$$

where

$$\phi = 0.65 \tag{76}$$

$$V_{bn} = \begin{cases} 2d_f t F_u & \text{for single shear connections} \\ 3d_f t F_u & \text{for double shear connections} \end{cases} \tag{77}$$

4. Bolt shear failure. The design shear force of bolts to ASTM standards shall be determined as,

$$V_{fvd} = \phi V_{fvn} \text{ (ASTM bolts)} \tag{78}$$

where

$$\phi = 0.65 \text{ (ASTM bolts)} \tag{79}$$

$$V_{fvn} = A_f F_{nv} \text{ (ASTM bolts)} \tag{80}$$

In equation (80), A_f is the bolt *shank* area and F_{nv} is the nominal shear strength, which is specified in the standards for shear planes through the shank and thread.

The Australian standard includes separate bolt shear design equations for bolts to ISO3506. The equations are based on those of Parts 1.1 and 1.4 of Eurocode3. According to the Australian standard, the design bolt shear force of bolts to ISO3506 shall be determined as,

$$V_{fvd} = \phi V_{fvn} \text{ (ISO 3506)} \tag{81}$$

where

$$\phi = 0.75 \text{ (ISO 3506)} \tag{82}$$

$$V_{fvn} = \begin{cases} 0.6A_f F_{nt} & \text{shear plane through shank} \\ 0.6A_{fs} F_{nt} & \text{shear plane through thread} \end{cases} \quad \text{(ISO 3506)} \tag{83}$$

In equation (83), A_{fs} is the tensile area of the bolt and F_{nt} is the nominal tensile strength.

(B) Tension. The design tensile force of bolts to ASTM standards shall be determined as,

$$N_{ftd} = \phi N_{ftn} \quad \text{(ASTM)} \tag{84}$$

where

$$\phi = 0.75 \quad \text{(ASTM)} \tag{85}$$

$$N_{ftn} = A_f F_{nt} \quad \text{(ASTM)} \tag{86}$$

In equation (86), A_f is the *shank* area and F_{nt} is the nominal reduced tensile strength which is specified in the Australian and American specifications. A reduced tensile strength is used because the strength is calculated on the shank area rather than the tensile area.

The specifications require that the pull-over capacity of the connected part be considered but does not provide design equations for this purpose.

For bolts to ISO3506, the Australian standard requires the design tensile force be determined as,

$$N_{ftd} = \phi N_{ftn} \quad \text{(ISO 3506)} \tag{87}$$

where

$$\phi = 0.67 \quad \text{(ISO 3506)} \tag{88}$$

$$N_{ftn} = A_{fs} F_{nt} \quad \text{(ISO 3506)} \tag{89}$$

In equation (89), A_{fs} is the tensile area and F_{nt} is the nominal tensile strength which is specified in the Australian Standard.

European standard.

According to Section 6 of Part 1.4 of Eurocode3, the design of bolted connections shall be to Section 6 of Part 1.1 of Eurocode3. The following two specific clauses are added:

- A separate check on the hole elongation under service loading can be avoided by using a reduced ultimate tensile strength calculated as,

$$F_{nt,red} = 0.5F_y + 0.6F_u \tag{90}$$

- The shear resistance for bolts to ISO3506 shall be determined as,

$$V_{fvd} = \frac{V_{fvn}}{\gamma_{Mb}} \tag{91}$$

where

$$\gamma_{Mb} = 1.35 \tag{92}$$

$$V_{fvn} = \begin{cases} 0.6A_f F_{nt} & \text{shear plane through shank} \\ 0.6A_{fs} F_{nt} & \text{shear plane through thread} \end{cases} \tag{93}$$

In equation (93), A_f and A_{fs} are the shank area and tensile area respectively and F_{nt} is the nominal tensile strength of the bolt. Equations (91-93) produce virtually the same design strength (V_{fvd}) as eqns (81-83).

Since Section 6.1 of Eurocode3, Part 1.4, only refers to Part 1.1 of Eurocode3, it is not clear whether the design provisions for connections of light-gauge structures covered by Part 1.3 of Eurocode3 can be used for cold-formed stainless steel structures. An exception is connections of stainless steel sheets using self-tapping screws, which according to Section 6.1 of Part 1.4 of Eurocode3 can be designed according to Part 1.3 of Eurocode3.

5.7.2.3 Design of welds

American and Australian/New Zealand specifications

The Australian and American provisions for welds are the same except that the Australian standard refers to AS/NZS1554.6 (1994) for prequalification and welding procedures, whereas the American specification refers to AWS D1.3 (1998). Design guidance is included for butt (groove) welds, fillet welds and resistance spot welds. The specifications do not contain design rules for arc spot (puddle) welds, arc seam welds or flare welds. The Australian standard contains guidance on the design of welded tubular connections.

The rules for butt and fillet welds are as follows:

(A). Butt welds. A distinction is made between butt welds subjected to tension (or compression) and shear, as shown in Figs **26**a and **26**b respectively

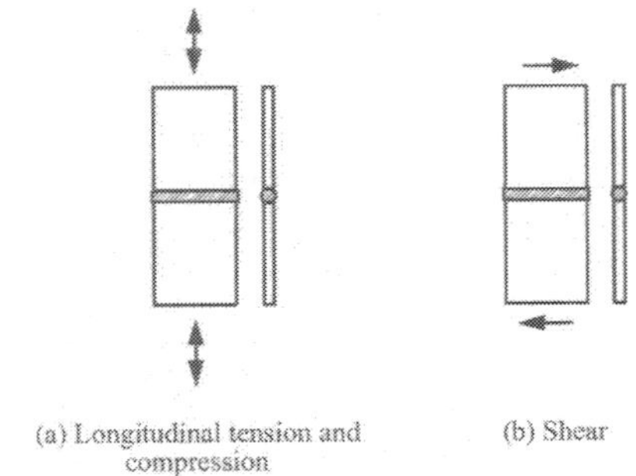

Figure 26. Butt welds in tension/compression and shear

1. Tension or compression. The design force in the connected part shall be determined as,

$$N_{wd} = \phi N_{wn} \tag{94}$$

where

$$\phi = 0.6 \tag{95}$$

$$N_{wn} = l_w t F_{ua} \tag{96}$$

In equation (96), l_w is the length of the weld, t is the thickness of the thinnest welded part, and (F_{ua}) is the *tensile* strength of the *annealed* base metal. The nominal strength equation (96) differs from equivalent equations for cold-formed and hot-rolled carbon steel structures, which use the yield stress rather than the tensile strength. Equation (96) originates from research at Cornell University (Errera et al. 1970) which showed good correlation between test strength and tensile strength. For 1/16, 1/4 and 1/2 hardened tempers, tests showed that the tensile strength of the annealed base metal should be used.

2. Shear. The design force in the connected part shall be determined as,

$$V_{wd} = \phi V_{wn} \tag{97}$$

where

$$\phi = 0.6 \tag{98}$$

$$V_{wn} = 0.6 l_w t F_{ua} \tag{99}$$

Equation (99) corresponds to eqn. (96) except that the approximate value of $0.6F_{ua}$ is used for the ultimate shear strength of the annealed base metal.

(B). Fillet welds. Design equations are provided for welds transferring loads by longitudinal and transverse shear, as shown in Figs **27**a and **27**b respectively.

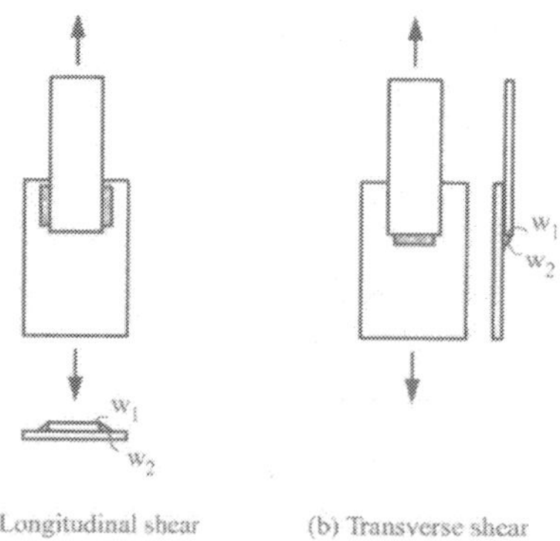

Figure 27. Fillet welds in longitudinal/transverse shear

1. Longitudinal loading. The design force in the connected part shall be determined as,

$$V_{wd} = \phi V_{wn} \tag{100}$$

where

$$\phi = 0.55 \tag{101}$$

$$V_{wn} = \min\{V_{wna}, V_{wnx}\} \tag{102}$$

$$V_{wna} = \begin{cases} (0.7 - 0.009 l_w / t) t l_w F_{ua} & \text{for } l_w / t < 30 \\ 0.43 t l_w F_{ua} & \text{for } l_w / t \geq 30 \end{cases} \tag{103}$$

$$V_{wnx} = 0.75 t_w l_w F_{xx} \tag{104}$$

In equation (104), t_w is the effective throat (min[$w_1/\sqrt{2}$, $w_2/\sqrt{2}$]) where w_1 and w_2 are the leg lengths), and F_{xx} is the tensile strength obtained from an all-weld-metal tensile test. Values for F_{xx} are given in the specifications for a wide range of electrodes.

2. Transverse loading. The design force in the connected part shall be determined as,

$$V_{wd} = \min\{\phi_a V_{wna}, \phi_x V_{wnx}\} \tag{105}$$

where

$$\phi_a = 0.55 \tag{106}$$

$$V_{wna} = t l_w F_{ua} \tag{107}$$

$$\phi_x = 0.65 \tag{108}$$

$$V_{wnx} = t_w l_w F_{xx} \tag{109}$$

European standard.

No specific design equations are given in Section 6.3 of Eurocode3, Part 1.4, for welded stainless steel connections. It is implicit that welded connections are covered by Section 6.1 General which refers to Part 1.1 of Eurocode3. As for bolted connections, there is no reference to Part 1.3 of Eurocode3 and one should assume that the design provisions of Part 1.3 do not apply to stainless steel welded connections.

5.7.3 Design of weld tubular connections

The Australian Standard contains design rules for welded joints in stainless steel tubes. The rules are based on tests reported in Rasmussen and Young (2001) and Rasmussen and Hasham (2001) on welded X- and K-joints in square and circular hollow sections (see Fig. **28**), which showed that the CIDECT design provisions (Wardenier et al. 1991; Packer et al. 1992) for welded carbon steel tubular joints can also be applied to stainless steel hollow sections. One of the significant conclusions was that in applying the CIDECT strength equations to cold-formed stainless steel tubes, the *enhanced* 0.2 % proof stress of the finished product may be substituted for the yield stress. For austenitic stainless steel tubes, the plastic deformations occurring during the cold-forming process typically double the 0.2 % proof stress, as mentioned in Section 5.3.2.

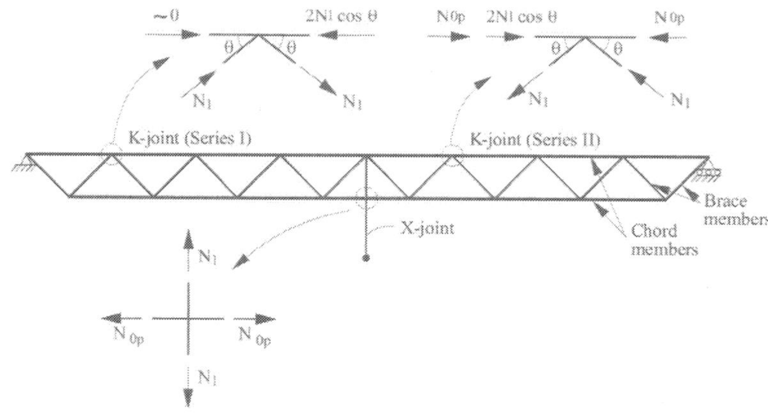

Figure 28. K- and X-joints in Warren truss

Appendix J of the Australian Standard contains the CIDECT strength equations in a similar form to that used in Annex K of Eurocode3, Part 1.1, (Eurocode3-1.1 1992). Unlike the CIDECT strength equations, which incorporate resistance factors in a non-explicit form, the strength equations given in Appendix J of the Australian Standard explicitly list the resistance factors. As an example, the table of strength equations for square hollow section chords welded to square or circular brace members (Table J.2 of AS/NZS4673 (2001)) is shown on Fig. **29**.

The range of applications covered by the CIDECT strength equations is significantly broader than that covered by the tests on stainless steel joints (Rasmussen and Hasham 2001; Rasmussen and Young 2001). However, the tests strengths were consistently conservative and it was deemed safe to extend the applicability range to that of the CIDECT strength equations on this basis.

The investigations described in Rasmussen and Young (2001) and Rasmussen & Hasham (2001) paid particular attention to the serviceability deformations of the joints. It was demonstrated that as a result of the gradual softening of stainless steel alloys, joint deformations grow at a faster rate than for carbon steel tubes leading to increased joint deformations at service loads. However, the serviceability deformation limit of 1 % of the chord width (or diameter) was not exceeded and thus it would not be necessary to check joint deformations under service loads when using the CIDECT strength equations. The deformation limit of 1 % of the chord width has emerged from the CIDECT research work as a *de facto* serviceability limit for welded tubular joints.

TABLE J7.2

DESIGN RESISTANCES OF WELDED JOINTS BETWEEN SQUARE OR CIRCULAR HOLLOW SECTION BRACE MEMBERS AND SQUARE HOLLOW SECTION CHORDS

Type of joint	Design resistance (i = 1 or 2, j = overlapped brace)
T, Y and X joints 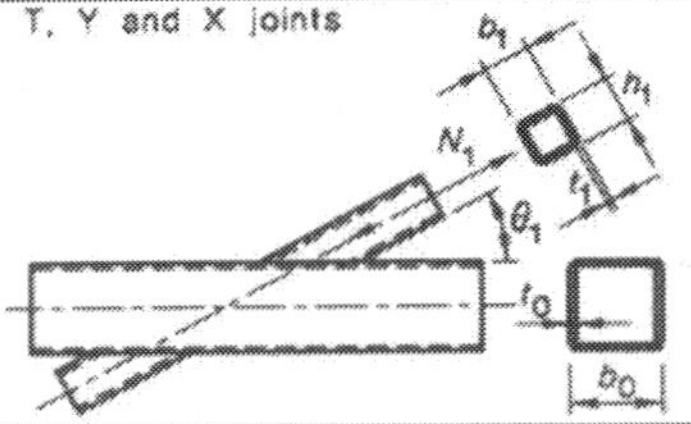	Chord face yielding $\beta \leq 0.85$ $\phi N_{1n} = \frac{f_{y0} t_0^2}{(1-\beta)\sin\theta_1}\left[\frac{2\beta}{\sin\theta_1} + 4(1-\beta)^{0.5}\right] k_n \left(\frac{\varphi}{0.9}\right)$
K and N gap joints 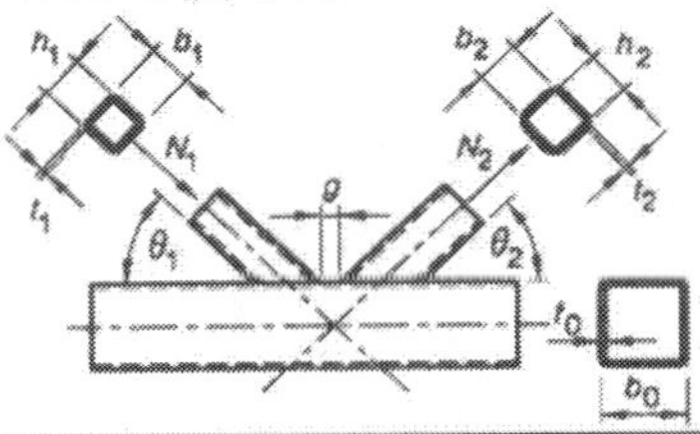	Chord face yielding $\beta \leq 1.0$ $\phi N_{in} = \frac{8.9 f_{y0} t_0^2}{\sin\theta_i}\left[\frac{b_1 - b_2}{2b_0}\right] \gamma^{0.5} k_n \left(\frac{\varphi}{0.9}\right)$
K and N overlap joints (see Note) 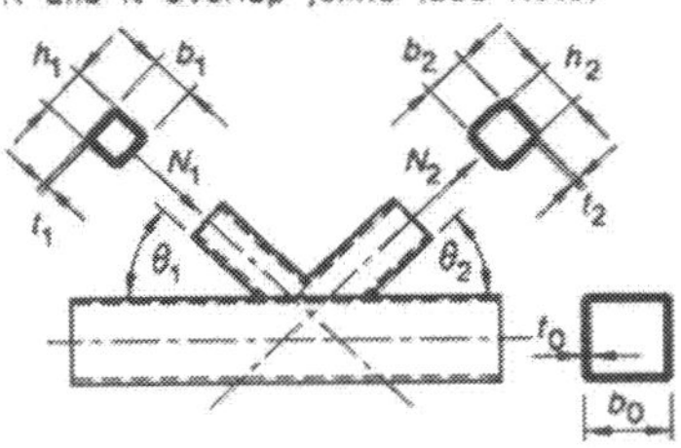	Effective width $25\% \leq \lambda_{ov} < 50\%$ $\phi N_{in} = f_{yi} t_i \left[\frac{\lambda_{ov}}{50}(2h_i - 4t_i) + b_{eff} + b_{e,ov}\right] \left(\frac{\varphi}{0.9}\right)$
	Effective width $50\% \leq \lambda_{ov} < 80\%$ $\phi N_{in} = f_{yi} t_i \left[2h_i - 4t_i + b_{eff} + b_{e,ov}\right] \left(\frac{\varphi}{0.9}\right)$
	Effective width $\lambda_{ov} \geq 80\%$ $\phi N_{in} = f_{yi} t_i \left[2h_i - 4t_i + b_i + b_{e,ov}\right] \left(\frac{\varphi}{0.9}\right)$
Circular braces	— Multiple the above design resistances by $\pi/4$. — Replace b_1 and h_1 with d_1. — Replace b_2 and h_2 with d_2.
Functions	
For $n \leq 0$ (tension): $k_n = 1.0$	For $n \geq 0$ (compression): $k_n = 1.3 - \frac{0.4n}{\beta}$ but $k_n \leq 1.0$
$b_{eff} = \left(\frac{10}{b_0/t_0}\right)\left(\frac{f_{y0} t_0}{f_{yi} t_i}\right) b_i$ but $b_{eff} \leq b_i$	$b_{e,ov} = \left(\frac{10}{b_j/t_j}\right)\left(\frac{f_{yj} t_j}{f_{yi} t_i}\right) b_i$ but $b_{e,ov} \leq b_i$

NOTE: Only the overlapping brace needs to be checked. The brace member efficiency, i.e. the design resistance of the joint divided by the design plastic resistance of the brace, for the overlapped brace should be taken as less than or equal to the overlapping brace.

Figure 29. Table J.2 of AS/NZS4673 (2001)

5.7.4 Design example, ultimate limit state capacity of single lap joint

Required: Determine the ultimate limit state design capacity of the lap joint shown in Figure 30 according to the Australian Standard AS/NZS4673. The bolts are M12 property class 70 to ISO 3506. The plate material is Duplex S31803.

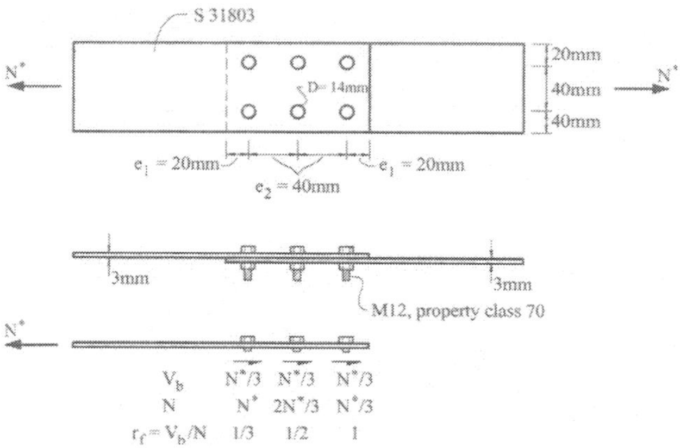

Figure 30. Single-shear lap joint

Solution:

Plate: The mechanical properties for S31803 alloy are given in Appendix B Table B1(A) of AS/NZS4673, and in Table **4** of Section 5.3.4,

$$F_{yt} = 430\,\text{MPa} \qquad F_{ut} = 590\,\text{MPa}$$

Bolt: The mechanical properties for M12 bolts, Property class 70, are given in Table 5.3.8 of AS/NZS4673,

$$F_{nt} = 700\,\text{MPa}$$

1. Tear-out capacity. Use eqns (65-67),

$$e = \min\{20\,\text{mm}, 40\,\text{mm}\} = 20\,\text{mm}$$

Nominal capacity of one bolt:

$$V_{fn} = teF_{ut} = 3 \times 20 \times 590\,\text{N} = 35.4\,\text{kN}$$

Tear-out design capacity of plate, (six bolts):

$$V_{fd} = 6 \times \phi V_{fn} = 6 \times 0.7 \times 35.4\,\text{kN} = 148\,\text{kN}$$

Check edge distances and hole spacing: The minimum edge distance (20 mm) exceeds $1.5d_f$=18 mm, and the hole spacing (40 mm) exceeds $3d_f$=36 mm, *ie* OK.

2. Net tension capacity. Use eqns (68-71).
Net area,

$$A_n = (b - 2d_h)t = (80 - 2 \times 14) \times 3 = 156\ \text{mm}^2$$

In eqn. (71), r_f is defined as the ratio between the force in the bolt(s) to the force in the plate at the bolt row considered, $r_f = V_b/N$. The values of r_f are calculated as 1/3, 1/2 and 1 for rows 1, 2 and 3 respectively, as shown in Fig. **30**. Also in eqn. (71), s_f=40mm.

First row of bolts:

$$\begin{aligned} F_{t_1} &= (1 - r_{f_1} + 2.5 r_{f_1} d_f / s_f F_u) \\ &= (1 - 1/3 + 2.5 \times (1/3) \times 12/40 \times 590 = 540\ \text{MPa} \end{aligned}$$

Second row of bolts:

$$\begin{aligned} F_{t_2} &= (1 - r_{f_2} + 2.5 r_{f_2} d_f / s_f F_u) \\ &= (1 - 1/2 + 2.5 \times (1/2) \times 12/40 \times 590 = 516\ \text{MPa} \end{aligned}$$

Third row of bolts:

$$\begin{aligned} F_{t_3} &= (1 - r_{f_3} + 2.5 r_{f_3} d_f / s_f F_u) \\ &= (1 - 1 + 2.5 \times 1 \times 12/40 \times 590 = 442\ \text{MPa} \end{aligned}$$

$$F_t = \min\{F_{t_1}, F_{t_2}, F_{t_3}\} = 442\ \text{MPa}$$

Nominal net tension capacity of plate:

$$N_{fn} = A_n F_t = 156 \times 422\ \text{N} = 65.8\ \text{kN}$$

Check general yielding

$$N_{fyn} = A_n F_{yt} = 156 \times 432\ \text{N} = 67.4\text{kN}$$

Net tension design capacity of plate:

$$N_{fyd} = \min\{\phi N_{fn}, \phi N_{fyn}\} = \min\{0.7 \times 65.8\ \text{kN}, 0.85 \times 67.4\ \text{kN}\} = 46.1\ \text{kN}$$

3. Bearing capacity. Use eqns (75-77).
Nominal bearing capacity of one bolt:

$$V_{bn} = 2 d_f t F_u = 2 \times 12 \times 3 \times 530\ \text{N} = 38.2\ \text{kN}$$

Design bearing capacity of plate, (six bolts):

$$V_{bd} = 6 \phi V_{bn} = 6 \times 0.65 \times 38.2\ \text{kN} = 166\ \text{kN}$$

4. Bolt shear capacity. Use eqns (81-83).
Nominal shear capacity of one bolt assuming fracture through the shank:

$$V_{fvn} = 0.6 \frac{\pi}{4} (d_f)^2 F_{nt} = 0.6 \times \frac{\pi}{4} \times (12)^2 \times 700\ \text{N} = 47.5\ \text{kN}$$

Design shear capacity of plate, (six bolts):

$$V_{fvd} = 6\phi V_{fvn} = 6 \times 0.75 \times 47.5\,\text{kN} = 214\,\text{kN}$$

The factored ultimate limit state design capacity of the lap joint can now be obtained as the minimum of the tear-out, net tension, bearing and shear capacities,

$$N_d = \min\{V_{fd}, N_{fd}, V_{bd}, V_{fvd}\} = \min\{148\,\text{kN}, 46.1\,\text{kN}, 166\,\text{kN}, 214\,\text{kN}\} = 46.1\,\text{kN}$$

The predicted failure mode is net section fracture.

References

AISI (1968a). Specification for the Design of Cold-formed Steel Structural Members, American Iron and Steel Institute, Washington, DC.

AISI (1968b). Specification for the Design of Light Gage Cold-formed Stainless Steel Structural Members, American Iron and Steel Institute, Washington, DC.

AISI (1974a). Specification for the Design of Cold-formed Stainless Steel Structural Members, American Iron and Steel Institute, Washington, DC.

AISI (1974b). Steel Products Manual - Stainless and Heat Resisting Steels, American Iron and Steel Institute, Washington, DC.

AISI (1986). Specification for the Design of Cold-formed Steel Structural Members, Washington, DC.

AISI (1991a). Design Guidelines for the Selection and Use of Stainless Steel, *A Designer's Handbook Series No. 9014*, American Iron and Steel Institute, Washington, DC.

AISI (1991b). Load and Resistance Factor Design Specification for Cold-formed Steel Structural Members, American Iron and Steel Institute, Washington D.C.

AISI (1997). Load and Resistance Factor Design Specification for Cold-formed Steel Structural Members, American Iron and Steel Institute, Washington, DC.

ANSI/ASCE-8 (1991). Specification for the Design of Cold-formed Stainless Steel Structural Members, American Society of Civil Engineers, New York, NY.

AS/NZS1554.6 (1994). Structural Steel Welding, Part 6: Welding Stainless Steels for Structural Purposes, AS/NZS1554.6, Standards Australia, Sydney.

AS/NZS4600 (1996). Cold-formed Structures, AS/NZS 4600, Standards Australia, Sydney.

AS/NZS4673 (2001). Cold-formed Stainless Steel Structures, AS/NZS4673, Standards Australia, Sydney.

ASSDA (2002). Australian Stainless Steel Reference Manual, Australian Stainless Steel Development Association, Brisbane.

ASTM (2002). Metals and Alloys in the Unified Numbering System, 9th ed, American Society for Testing and Materials, Philadelphia.

ASTM-A176 (1999). Standard Specification for Stainless and Heat-Resisting Chromium Steel Plate, Sheet, and Strip, A176, American Society for Testing and Materials, Philadelphia.

ASTM-A193 (2001). Standard Specification for Alloy-Steel and Stainless Steel Bolting Materials for High-Temperature Service, A193, American Society for Testing and Materials, Miami.

ASTM-A240 (2002). Standard Specification for Chromium and Chromium-Nickel Stainless Steel Plate, Sheet, and Strip for Pressure Vessels and for General Applications, A240, American Society for Testing and Materials, Philadelphia.

ASTM-A276 (2000). Standard Specification for Stainless Steel Bars and Shapes, A276, American Society for Testing and Materials, Philadelphia.

ASTM-A666 (2000). Standard Specification for Annealed or Cold-Worked Austenitic Stainless Steel Sheet, Strip, Plate, and Flat Bar, A666, American Society for Testing and Materials, Philadelphia.

ASTM-F593 (2001). Standard Specification for Stainless Steel Bolts, Hex Cap Screws, and Studs, F593, American Society for Testing and Materials.

AWS-C1.1 (2000). Recommended Practices for Resistance Welding, C1.1, American Welding Society, Miami.

AWS-D1.3 (1998). Structural Welding Code - Sheet Steel, D1.3, American Welding Society, Miami.

Baddoo, N. and B. Burgan (2001). Structural Design of Stainless Steel, Steel Construction Institute, London.

Berg, G. v. d. (2000). "The Effect of the Non-linear Stress-strain Behaviour of Stainless Steels on Member Capacity." *Journal of Constructional Steel Research* **54**(1): 135-160.

Bleich, F. (1952). *Buckling Strength of Metal Structures*. New York, NY, McGraw-Hill.

BSI (2001). Structural use of Steelwork in Building. Specification for Materials, Fabrication and Erection: Hot-Rolled Sections, British Standards Institution, London.

Burgan, B. (1993). Concise Guide to the Structural Design of Stainless Steel, The Steel Construction Institute, Ascot, UK.

Chryssantholopoulos, M. and Y. Low (2001). "A Method for Predicting the Flexural Response of Tubular Members with Non-linear Stress-strain Characteristics." *Journal of Constructional Steel Research* **57**(11): 1197-1216.

Clarke, M. (1994). Plastic-zone Analysis of Frames. *Advanced Analysis of Steel Frames: Theory, Software and Applications*. W. C. a. S. Toma. London, CRC Press: Chapter 6.

CMC (1966). Corrosion in Action, Climax Molybdenum Company, Greenwich, Conn.

EN-10088 (1995). Stainless Steels, EN-10088, European Committee for Standardisation, Brussels.

Engesser, F. (1889). *Zeitschrift fur Architektur und Ingenieurwesen* **35**: 455.

Engesser, F. (1895). *Schweizerische Bauzeitung* **26**: 24.

Errera, S., B. Tang and D. Popovich (1970). Strength of Bolted and Welded Connections in Stainless Steel, *Report No. 335*, Department of Civil Engineering, Cornell University, Ithaca, NY.

Eurocode3-1.1 (1992). Eurocode3: Design of Steel Structures, Part 1.1: General Rules and Rules for Buildings, ENV-1993-1-1, European Committee for Standardisation, Brussels.

Eurocode3-1.3 (1996). Eurocode3: Design of Steel Structures, Part 1.3: Cold Formed Thin Gauge members and Sheeting, ENV-93-1-3, European Committee for Standardisation, Brussels.

Eurocode3-1.4 (1996). Eurocode3: Design of Steel Structures, Part 1.4: Supplementary Rules for Stainless Steel, prENV-93-1-4, European Committee for Standardisation, Brussels.

Gerard, G. (1946). "Secant Modulus Method for Determining Plate Instability above the Proportionality Limit." *Journal of the Aeronautical Sciences* **13**: 38.

Hibbitt, Karlsson and Sorensen (1995). *ABAQUS Standard, Users Manual, Ver. 5.5*.

ISO-3506 (1997). Mechanical Properties of Corrosion-resistant Stainless-steel Fasteners, ISO 3506, International Standards Organisation, Geneva.

Johnson, A. and G. Winter (1966). "Behaviour of Stainless Steel Columns and Beams." *Journal of Structural Engineering, American Society of Civil Engineers, ASCE* **92**(ST5): 97-118.

Lin, S.-H., W.-W. Yu and T. Galambos (1988a). Design of Cold-formed Stainless Steel Structural Members, Proposed Allowable Stress Design Specification with Commentary, *Progress Report No. 3*, University of Missouri-Rolla, Rolla, MO.

Lin, S.-H., W.-W. Yu and T. Galambos (1988b). Load and Resistance Factor Design of Cold-formed Stainless Steel, Statistical Analysis of Material Provisions and Development of the LRFD Provisions, *Progress Report No. 4*, University of Missouri-Rolla, Rolla, MO.

Lula, R. (1965). Stainless Steel, American Society of Metals, Ohio.

Mann, A. (1993). "The Structural Use of Stainless Steel." *The Structural Engineer* **71**(4): 60-69.

Mirambell, E. and E. Real (2000). "On the calculation of deflections in structural stainless steel beams: an experimental and numerical investigation." *Journal of Constructional Steel Research* **54**(1): 109-133.

NiDI (1990). Advantages for Architects: An Architect's Guide on Corrosion Resistance, Nickel Development Institute, Toronto.

NiDI (1992). Guidelines for the Welded Fabrication of Nickel-containing Stainless Steels for Corrosion Resistant Services, Nickel Development Institute, Toronto.

Packer, J., J. Wardenier, Y. Kurobane, D. Dutta and N. Yeomans (1992). Design Guide for Rectangular Hollow Section (RHS) Joints under Predominantly Static Loading, Comite International pour le Developpement et l'Etude de la Construction Tubulaire (CIDECT), Verlag TUV Rheinland, Cologne.

Peckner, D. and I. Bernstein (1977). *Handbook of Stainless Steels*. New York, N.Y., McGraw-Hill.

Ramberg, W. and W. Osgood (1943). Description of Stress Strain Curves by Three Parameters, *Technical Note No. 902*, National Advisory Committee for Aeronautics, Washington, DC.

Rasmussen, K. (2001). Full-range Stress-strain Curves for Stainless Steel Alloys, *Research Report R811*, Department of Civil Engineering, University of Sydney.

Rasmussen, K., P. Bezkorovainy and T. Burns (2002). Strength Curves for Metal Plates, *Research Report*, Department of Civil Engineering, University of Sydney.

Rasmussen, K. and G. Hancock (1992). *Design of Cold-formed Stainless Steel Tubular Beams*, Recent Developments in Cold-formed Steel Design and Construction, 11th International Specialty Conference on Cold-formed Steel Structures, Ed(s) W.-W. Yu and R. LaBoube, St Louis, University of Missouri-Rolla.

Rasmussen, K. and G. Hancock (1993a). "Stainless Steel Tubular Members. I: Columns." *Journal of Structural Engineering, American Society of Civil Engineers, ASCE* **119**(8): 2349-2367.

Rasmussen, K. and G. Hancock (1993b). "Stainless Steel Tubular Members. II: Beams." *Journal of Structural Engineering, American Society of Civil Engineers, ASCE* **119**(8): 2368-2386.

Rasmussen, K. and A. Hasham (2001). "Tests of X- and K-joints in CHS Stainless Steel Tubes." *Journal of Structural Engineering, American Society of Civil Engineers* **127**(10): 1183-1189.

Rasmussen, K. and J. Rondal (1997). "Strength Curves for Metal Columns." *Journal of Structural Engineering, ASCE* **123**(6): 721-728.

Rasmussen, K. and J. Rondal (2000). "Column Curves for Stainless Steel Alloys." *Journal of Constructional Steel Research* **54**(1): 89-107.

Rasmussen, K. and B. Young (2001). "Tests of X- and K-joints in SHS Stainless Steel Tubes." *Journal of Structural Engineering, American Society of Civil Engineers* **127**(10): 1173-1182.

SABS-0162-4 (1997). Structural Use of Steel, Part4: The design of Cold-formed Stainless Steel Structural Members, SABS 0162-4, South African Bureau of Standards, Pretoria.

Sedriks, A. (1979). *Corrosion of Stainless Steels*. New York, N.Y., John Wiley and sons.

Shanley, F. (1947). "Inelastic Column Theory." *Journal of the Aeronautical Sciences* **14**(5): 261-267.

Simmons, W. and J. v. Echo (1965). The Elevated-temperature Properties of Stainless Steel, *ASTM Data Series Publication DS5-S1*, American Society for Testing and Materials, Philadelphia.

Stowell, E. (1948). A Unified Theory of Plastic Buckling of Columns and Plates, *Technical Note No. 1556*, National Advisory Committee for Aeronautics, Washington, DC.

Talja, A. and P. Salmi (1995). Design of Stainless Steel RHS Beams, Columns and Beam-columns, *Research Notes 1619*, Technical Research Centre of Finland, VTT, Espoo.

Wardenier, J., Y. Kurobane, J. Packer, D. Dutta and N. Yeomans (1991). Design Guide for Circular Hollow Section (CHS) Joints under Predominantly Static Loading, Comite International pour le Developpement et l'Etude de la Construction Tubulaire (CIDECT), Verlag TUV Rheinland, Cologne.

Chapter 6: High Strength Steel Structures

K.J.R. Rasmussen

Department of Civil Engineering, University of Sydney, Sydney, Australia
E-mail: K.Rasmussen@civil.usyd.edu.au

6.1 Introduction

The efficiency of steel structural members and connections can in many cases be enhanced by using steels with high values of yield stress and/or tensile strength. Good economy can be achieved by utilizing the superior strength-to-weight ratio of such steels, particularly in relatively heavy construction. In light gauge construction, high strength steels offer ease of handling and quick construction.

The mechanical properties, notably strength and ductility, of steel depend primarily on its microstructure, that is the arrangement and chemical composition of the microscopic crystals of which the steel is composed. The microstructure depends mainly on:

- chemical composition,
- thermal history and
- work-hardening history,

and can be changed greatly by changing any of these influencing factors.

With respect to chemical composition, the mechanical properties are most significantly influenced by the carbon content. The strength increases with increasing carbon content but the weldability decreases. If the steel is to be weldable, the permitted range of carbon content is relatively limited and consequently, the most common processes for enhancing the mechanical properties of steel are by changing the thermal history and/or the work-hardening history.

In relatively heavy construction, say plate thickness of 5 mm or larger, the most common process for enhancing strength is by *quenching and tempering*. This is a thermal process in which the steel is first cooled rapidly (quenched) to achieve high strength and subsequently heated (tempered) to partially anneal the steel and gain ductility and weldability. The most common grades of structural quenched and tempered steels have yield stress values in the range from 420 MPa to 690 MPa. The process for producing quenched and tempered steels and the design of such steels are described further in Section 6.2.

Cold-reduced light-gauge steels fall in the category of work-hardened steels. These steels have typical thicknesses less than 3 mm and yield stress values in the range from 300 MPa to 550 MPa depending mainly on thickness. They gain increased strength by the plastic deformations occurring during the thickness-reducing rolling process. The production of these steels and the design of such steels are briefly described in Section 6.3.

6.2 Quenched and Tempered High Strength Steels

6.2.1 General

Quenched and tempered steels are typically used in long span applications, such as bridges, where it is desirable to minimize the self-weight. Because of their relatively high cost, they may be combined with ordinary steel to form hybrid sections, such as deep I-girders with high strength steel flanges and ordinary steel webs.

The most common American and Australian quenched and tempered structural steels conform to ASTM standards A852 (2001) and A514 (2000). The European standard covering quenched and tempered structural steels is EN10113 (1997).

6.2.2 The Quenching and Tempering Process

The metallurgical characteristics of constructional alloy steels were described by Murphy (1964). The following excerpt is useful in explaining specific procedures required in the design and fabrication of quenched and tempered steels.

The fundamental property of steel, from which its response to heat treatment derives, is the ability to exist with its atoms arranged in two distinctly different crystallographic forms: one which is characteristic of steel at high temperatures (above 830°C for a 0.2% carbon content) and one which is characteristic of steel at lower temperatures. The higher temperature crystallographic form of steel, called austenite, can dissolve nearly 2% carbon in solid solution, whereas the low-temperature crystallographic form, called ferrite, can hold no more than about 0.001% carbon in solid solution at room temperature. The excess carbon is rejected as the intermetallic compound iron carbide. The microstructure thus produced consists of ferrite and lamellar mixture of iron carbide and ferrite called pearlite.

Ferrite is a relatively low strength but relatively tough and ductile material whereas iron carbide is an extremely strong but brittle material. Fortunately, mixtures of these two constituents, in general, combine the properties of both to produce a relatively strong and yet ductile and tough material. Non-heat-treated weldable steels used for general structural applications generally exhibit ferrite-pearlite microstructures and have a yield stress (or 0.2% proof stress) in the range from 200 MPa to 300 MPa. The toughness of these steels is generally adequate as indicated by their widespread use.

The microstructure of steel can be changed radically by increasing the rate of cooling from the austenising temperature. Cooling of the 0.2% carbon steel from the austenising temperature of 930°C to room temperature at an extremely rapid rate results in a yield stress of about 1000 MPa, an increase of about 750 MPa over that obtained as a result of slow cooling to room temperature. The increase in strength is a result of the transformation of the austenite to a microstructure completely different from the ferrite-pearlite microstructure observed previously. Because the transformation from austenite to the high temperature transformation products is time dependent, high cooling rates prevent the transformation to pearlite. In addition, the temperature of transformation of austenite to ferrite is suppressed by relatively rapid cooling rates. The microstructural constituents resulting from the rapid cooling of steel are, therefore, generally called low temperature transformation products. One of these microconstituents is martensite, and is generally considered to be the strongest of the microstructures of steel. Martensite forms during cooling from the austenite when the cooling rate is sufficiently rapid so that there is not sufficient time for rejection of carbon as iron carbide. The result is a super-saturated solid solution of carbon in ferrite. Another microconstituent that can be formed as a result of rapid cooling is the low-temperature transformation product called bainite.

The primary function of alloying elements in steel is to increase the ease with which low temperature transformation products such as martensite can be obtained when slower and slower cooling rates are used. The property of steel that refers to its capacity to be hardened, that is to form low temperature transformation products is called hardenability. It should be noticed that hardness and hardenability are not synonymous; hardenability refers to the ease with which a specific level of hardness may be obtained in a steel.

The heat-treating process whereby steels are hardened or strengthened consists of three separate steeps:

1. The steel is heated to a temperature at which it is completely austenitic and at which the carbon (and alloying elements) can dissolve in solid solution.
2. The steel is held at the austenising temperature for a time sufficient to ensure uniformity of temperature through the section thickness and to ensure complete solution of the carbon and alloying elements.
3. The steel is cooled at a rate that will result in high percentages of low temperature transformation products such as martensite. The cooling rate depends on the alloy and carbon content of the steel. However, for 0.1 to 0.25% carbon constructional steels, water quenching is almost invariably used.

Although steels that have been quenched to obtain martensitic microstructures exhibit high strength, they do generally not exhibit adequate toughness and ductility for use in engineering structures. Therefore, after quenching they are subjected to an additional heat treatment called tempering. The tempering process involves heating the steel to a temperature below which the austenite begins to re-form (at about 700°C) and holding at this temperature for a specific period of time before cooling to room temperature. At the tempering temperature, the carbon atoms present are thermally activated and can diffuse readily through the ferrite. As a result, the carbon in the martensite precipitates from solid solution and forms a uniform dense dispersion of iron and alloy carbides in a ferrite matrix. Carbide precipitation during tempering produces a reduction in strength and an increase in toughness and ductility. Tempering times and temperatures can be chosen so that a desired combination of strength and toughness may be obtained.

6.2.3 The Stress-strain Curve

Figure **1** shows typical stress-strain curves obtained from tension and compression coupon tests on 6 mm thick Grade 690 quenched and tempered steel (Rasmussen and Hancock 1992). The following observations can be made, which apply generally to quenched and tempered steels:

- The stress-strain curves do not have upper and lower yield stress values as for ordinary Grade 250 MPa and 350 MPa steel.
- Gradual yielding takes place near the yield plateau, and accordingly, the yield stress is usually defined as the 0.2% proof stress.

- The uniform elongation corresponding to the ultimate tensile strength is of the order of 8 %.
- The ratio of ultimate tensile strength to tensile yield stress is of the order of 1.1.

The ratio of ultimate tensile strength to tensile yield stress is significantly lower than for ordinary steel grades for which it is about 1.4. Furthermore, the ductility is significantly less than for ordinary steel grades for which is it about 30%. Thus, as the yield stress is increased, the ductility and the capacity to strain harden gradually decrease. The reduction in ductility and strain hardening capacity have ramifications for the structural design of quenched and tempered steels, as described in Sections 6.2.4 and 6.2.5 following. The welding of quenched and tempered steels also requires added care in order to maintain the high percentage of martensite in the crystal structure, and avoid weld-metal cracking in restrained welds. Preheat is often used to overcome these welding problems.

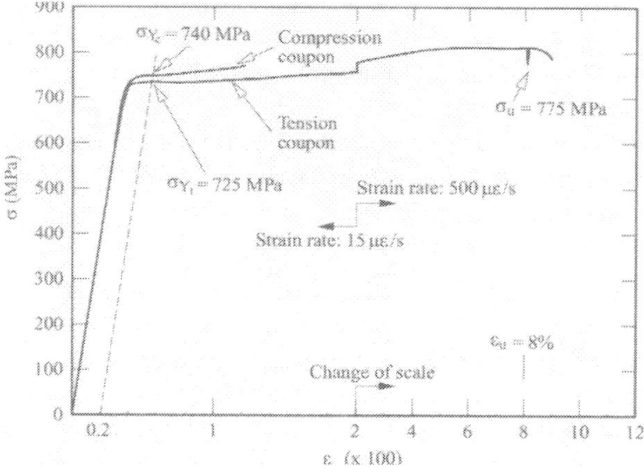

Figure 1. Typical stress-strain curves of Grade 690 quenched and tempered steel, (Rasmussen and Hancock 1992)

6.2.4 Design of compression members

6.2.4.1 Section capacity, (design for local buckling)

In the stocky slenderness range where plate elements reach yield before failing, the strength is affected by the strain hardening capacity. It is well-known from early tests at Cambridge University (Moxham 1971), that stocky plates may reach ultimate average stresses up to 30% higher than the yield stress. Hence, high strength steels can be expected to have relatively low strength in the stocky slenderness range. At the same time, welding residual stresses may have less influence on the strength of high strength steels compared to ordinary grades because the magnitudes of compressive residual stress found in welded high strength sections do not differ greatly from those found in ordinary steel sections, so that the ratio of compressive residual stress to yield stress may be lower for high strength steels than for ordinary steels. It is the ratio of compressive residual stress to yield stress, rather than the magnitude of compressive residual stress itself, which determines the reduction in strength. For these reasons, it can be expected that the plate strength curves for high strength steels are different from those for ordinary steel.

Numerous tests have been conducted to determine strength curves for high strength steel plates (Fukumoto and Itoh 1984). Some of these tests (Nishino et al. 1966; Usami and Fukumoto 1982) and more recent tests (Rasmussen and Hancock 1992) on stiffened and unstiffened elements are shown in Figures 2a and 2b respectively. The solid markers in these figures are for tests on 690 MPa quenched and tempered steels, and the open markers and numerals are for Cambridge University tests on ordinary steel plates.

As shown by the test points in Figs 2a and 2b, the nondimensional plate strength of high strength steel is generally lower in the stocky slenderness range, say $\lambda<0.7$, as could be expected because of the reduced strain-hardening capacity. In the high slenderness range, the tests on unstiffened elements suggest that the nondimensional plate strength of high strength steel may be higher, as shown in Fig. 2b. There may also be a slight increase in strength in the high slenderness range for stiffened elements, as shown in Fig. 2a, but the increase is less pronounced than for unstiffened elements. The general conclusion emerges from Figs 2a and 2b that the plate strength curves for high strength steels are flatter than those for ordinary steels and have approximately the same yield slenderness limit, *ie* value of slenderness (λ_e) beyond which the plate strength is reduced below the yield stress as a result of local buckling.

Figures **2**a and **2**b also compare plate test results with the main international standards for hot-rolled steel structures, including Eurocode3, Part 1.1, (Eurocode3-1.1 1992), the American Institute of Steel Construction Load and Resistance Factor Specification (AISC-LRFD 1999) and the Australian Standard AS4100 (1998). It follows from Fig. **2**a that the Eurocode3 and AISC-LFRD strength curves for stiffened elements are optimistic compared to the tests on Grade 690 high strength steel plates, particularly at large slenderness values. The AISC-LRFD strength curve for unstiffened elements is slightly optimistic while Eurocode3 is conservative compared to the test strengths. The Australian Standard (AS4100) is generally conservative compared to the tests on high strength steel plates, except near the yield slenderness limit.

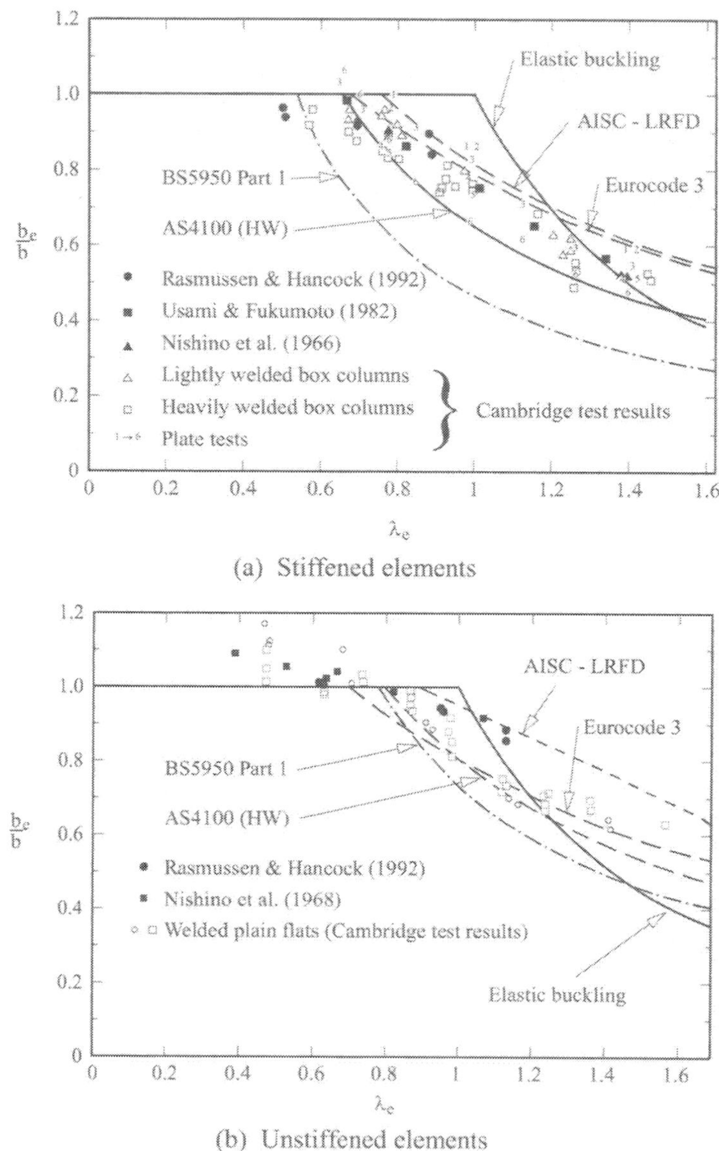

Figure 2. Test strengths and strength curves for plate elements (Rasmussen and Hancock 1992)

6.2.4.2 Effect of residual stresses

As mentioned earlier, high strength steel sections are less affected by residual stresses than ordinary steel sections because the ratio of residual stress to yield is less for high strength steel plates. The following simple analysis further explains this reasoning:

Denoting the local buckling and yield stresses by σ_{cr} and σ_y respectively, the plate strength (σ_u) may be determined using the von Karman formula,

$$\frac{\sigma_u}{\sigma_y} = \frac{1}{\sqrt{\frac{\sigma_y}{\sigma_{cr}}}}, \quad \sqrt{\frac{\sigma_y}{\sigma_{cr}}} \geq 1 \tag{1}$$

This expression assumes that the plate contains neither geometric imperfections nor residual stresses. If a compressive residual stress (σ_r) is present, the local buckling stress is approximately given by,

$$\sigma_{cr\,r} = \sigma_{cr} - \sigma_r \tag{2}$$

Hence, (not accounting for the effect of geometric imperfections), the strength is obtained as

$$\frac{\sigma_u}{\sigma_y} = \frac{1}{\sqrt{\frac{\sigma_y}{\sigma_{cr_r}}}} = \frac{\alpha}{\sqrt{\frac{\sigma_y}{\sigma_{cr}}}}, \quad \sqrt{\frac{\sigma_y}{\sigma_{cr}}} \geq \alpha \tag{3}$$

where the reduction factor (α) is given by,

$$\alpha = \sqrt{1 - \frac{\sigma_r}{\sigma_y}\frac{\sigma_y}{\sigma_{cr}}} = \sqrt{1 - \frac{\sigma_r}{\sigma_y}\lambda_e^2} \tag{4}$$

$$\lambda_e = \sqrt{\frac{\sigma_y}{\sigma_{cr}}} \tag{5}$$

$$\sigma_{cr} = \frac{k\pi^2 E}{12(1-\nu^2)}\left(\frac{t}{b}\right)^2 \tag{6}$$

According to eqns (3,4), the reduction factor (α) and hence the plate strength decrease with increasing value of σ_r/σ_y. Furthermore, the reduction factor (α) decreases with increasing slenderness (λ_e). Consequently, the difference in nondimensional strength of ordinary and high strength steel plates should increase as the slenderness increases. This conclusion was also drawn by Nishino et al. (1966).

However, in drawing these conclusions it is tacitly assumed that the residual stress (σ_r) is the same for ordinary and high strength steel plates of a certain slenderness (λ_e). This assumption is generally not correct as may be seen by considering an ordinary and a high strength steel plate of the same slenderness (λ_e), as shown in Fig. **3**. The superscrips "o" and "h" denote "ordinary" and "high strength" respectively. When the plates are assumed to have the same slenderness and have the same thickness, their widths are related by,

$$b^h = \sqrt{\frac{\sigma_y^o}{\sigma_y^h}}\, b^o \tag{7}$$

The plate areas (A=bt) are similarly related by,

$$A^h = \sqrt{\frac{\sigma_y^o}{\sigma_y^h}}\, A^o \tag{8}$$

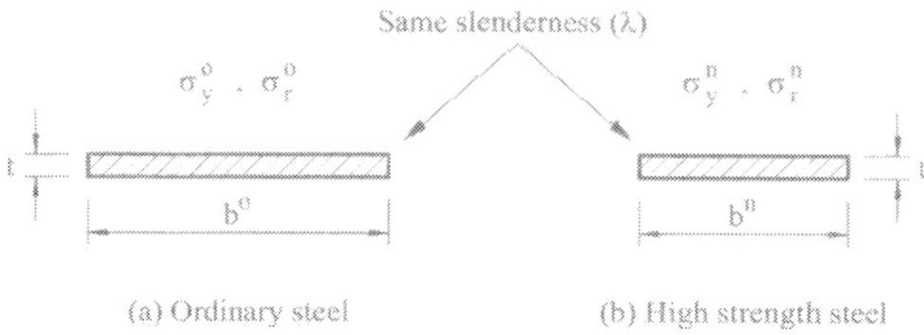

Figure 3. Ordinary and high strength steel plate of same slenderness.

The resultant of the compressive residual stress ($\sigma_r A$) is proportional to the heat put into the plate during welding. The heat input depends on the voltage, amps and welding speed, as well as the number of weld runs. It is therefore a function of thickness while is can be assumed to be independent of plate width. Thus, the heat input can be assumed to be equal for the two plates of equal thickness shown in Fig. 3. It follows that the compressive residual stresses in the two plates are related by,

$$\sigma_r^h = \frac{A^o}{A^h}\sigma_r^o \tag{9}$$

Combining eqns (8,9), the residual stresses in the two plates are related as follows,

$$\sigma_r^h = \sqrt{\frac{\sigma_y^h}{\sigma_y^o}}\sigma_r^o \tag{10}$$

Using eqn. (4), the reduction factors (α) for the two plates are,

$$\alpha^o = \sqrt{1 - \frac{\sigma_r^o}{\sigma_y^o}\lambda_e^2} \tag{11}$$

$$\begin{aligned} \alpha^h &= \sqrt{1 - \frac{\sigma_r^h}{\sigma_y^h}\lambda_e^2} \\ &= \sqrt{1 - \frac{\sigma_r^o}{\sigma_y^h}\sqrt{\frac{\sigma_y^h}{\sigma_y^o}}\lambda_e^2} \\ &= \sqrt{1 - \frac{\sigma_r^o}{\sigma_y^o}\lambda_e^2\sqrt{\frac{\sigma_y^o}{\sigma_y^h}}} \end{aligned} \tag{12}$$

Combining eqns (11,12), the relationship between the reduction factors (α) for ordinary and high strength steel plates is obtained as,

$$\alpha^h = \sqrt{1 - \sqrt{\frac{\sigma_y^o}{\sigma_y^h}}\left[1 - \left(\alpha^o\right)^2\right]} \tag{13}$$

By substituting the following typical values ($\sigma_y^h = 690\,\text{MPa}$, $\sigma_y^o = 300\,\text{MPa}$, α^o=0.7) into eqn. (13), the reduction factor (α^h) for high strength steel is obtained as 0.81, or 16 % higher than for ordinary steel.

So far, the effect of residual stresses on *plate* strength has been discussed and equations have been derived to qualitatively assess differences in strength of ordinary and high strength steel plates. However, residual stresses also affect the strength of members, notably columns and unbraced beams for which tensile and compressive residual stresses precipitate yielding in the parts of the cross-section subjected to tension and compression respectively. In the case of columns, the member slenderness $\lambda=\sqrt{\sigma_y/\sigma_E}$ is a function of the yield stress and the Euler buckling stress,

$$\sigma_E = \frac{\pi^2 E}{(L_e / r)^2} \tag{14}$$

where L_e is the effective length, which depends on the end support conditions, and r is the radius of gyration,

$$r = \sqrt{\frac{I}{A}} \tag{15}$$

The Euler buckling stress is independent of the yield stress, and so an ordinary and a high strength steel column with the same cross-section and same support conditions will have the same Euler buckling stress. They will also have approximately the same level of compressive residual stress because the heat input would be approximately the same in the two columns. Thus, if one considers two geometrically identical columns, one fabricated from ordinary steel and one from high strength steel, then the residual stress to yield stress ratio will be significantly lower for the high strength steel column and hence its nondimensional strength should be higher.

6.2.4.3 Member capacity, (flexural buckling)

To investigate the strength of high strength steel columns, Nishino and Tall (1970) performed tests on rolled and welded high strength ASTM-514 steel columns with a nominal yield stress of 690 MPa. Numerical studies on high strength steel columns were carried out in Europe at about the same time (ECCS 1977). Both studies suggested that high strength steel columns may be designed to a higher column curve than ordinary steel columns when compared on a nondimensional basis. However, in the American tests, the columns were loaded concentrically in accordance with the American design philosophy which was to base the design strength on straight concentrically loaded columns and allow for overall imperfections by using a relatively small resistance factor. The positions of the loading points were adjusted so that the effects of geometric imperfections and loading eccentricity approximately counteracted each other, leading to nearly perfect bifurcation behaviour.

Further tests were conducted by Rasmussen and Hancock (1995) for determining the strength of high strength steel columns with initial crookedness. Tests were performed on box and I-sections fabricated from nominally 690 MPa steel plates. The sections were designed to reach yield before local buckling. Long column tests were performed between pinned ends at three lengths, and short (stub column) tests were performed between fixed ends. Two tests were performed at each long column length. In the first, the column was nominally concentrically loaded whereas in the second, the load was applied with a nominal eccentricity of $L/1000$, where L is the pin-ended column length. This was the maximum out-of-straightness permitted of a column in the Australian steel structures standard AS4100 (1998). The main purpose of the tests was to use the *eccentric* tests to select an appropriate column curve for high strength steel columns. The I-sections buckled about their minor axis.

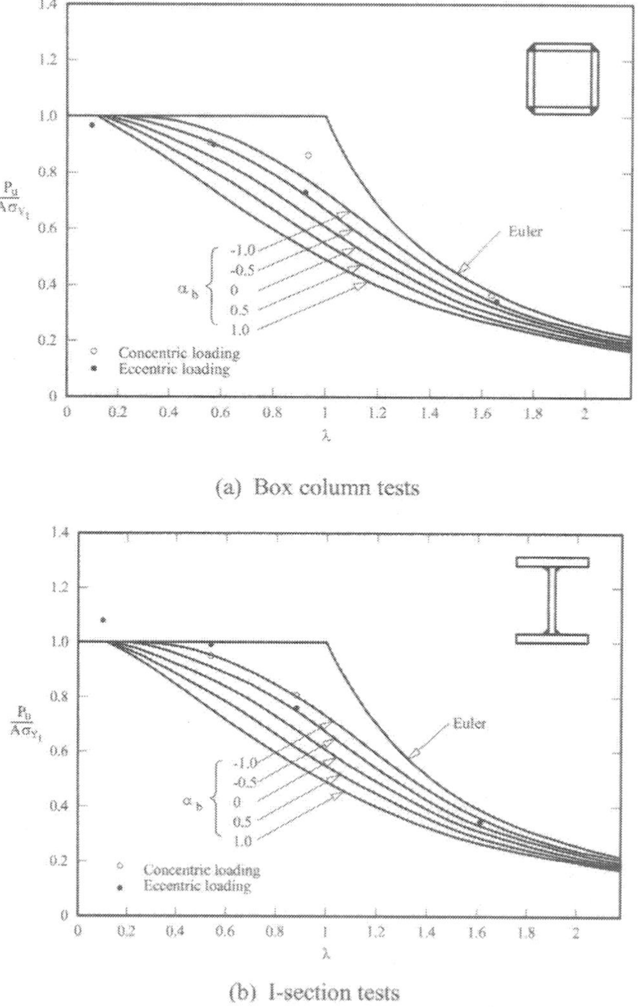

(a) Box column tests

(b) I-section tests

Figure 4. Test strengths and Australian strength curves for Grade 690 MPa long columns (Rasmussen and Hancock 1995)

The tests results are shown in Figs **4**a and **4**b for box and I-section columns respectively. The column strengths are nondimensionalised with respect to the squash load ($A\sigma_y$) and the slenderness (λ) is defined as,

$$\lambda = \sqrt{\frac{\sigma_y}{\sigma_E}} \tag{16}$$

where σ_E is the Euler buckling stress, (see eqn. (20) following). The test strengths are compared with the five column curves included in the Australian standard, defined by the parameter α_b which takes the values -1, -0.5, 0, 0.5 and 1. For box and I-sections fabricated from ordinary steel grades, the appropriate column curve is the α_b=0 curve. However, it follows from Figs **4**a and **4**b that the α_b=-0.5 curve is appropriate for 690 MPa steel columns, as shown by the correspondence between this curve and the tests on eccentrically loaded columns.

The box and I-section test strengths are compared in Figs 5a and 5b respectively with the column strength curves specified in the American Institute of Steel Construction Load and Resistance Specification (AISC-LRFD 1999), the British Standard BS5950, Part 1, (BSI 2001) Eurocode3, Part 1.1, (Eurocode3-1.1 1992) and the Australian steel structures standard AS4100 (1998). The Eurocode3 curves are based on Section 5.5.1 of the specification. Annex D of Eurocode3 allows higher design curves to be used for *hot-rolled* Grade 420 MPa and 460 MPa I-sections compared to hot-rolled I-sections of ordinary Grade 225 MPa, 275 MPa and 355 MPa European steel. However, the strength curves nominated in Annex D for *welded* Grade 420 MPa and 460 MPa box and I-sections are the same as those given in Section 5.5.1 of Part 1.1 for ordinary steel grades.

The Eurocode3 adoption of the Rondal-Maquoi approximations (Rondal and Maquoi 1979) to the multiple ECCS (1977) "a", "b" and "c" column curves is used as reference in the following discussion of the strength curves shown in Figs 5a and 5b.

According to Rondal and Maquoi (1979), the ECCS "a", "b" and "c" curves are approximated closely by the slenderness reduction factor,

$$\chi = \frac{1}{\varphi + \sqrt{\varphi^2 - \lambda^2}} \le 1 \tag{17}$$

where

$$\varphi = \tfrac{1}{2}(1 + \eta + \lambda^2) \tag{18}$$

$$\lambda = \sqrt{\frac{\sigma_y}{\sigma_E}} \tag{19}$$

$$\sigma_E = \frac{\pi^2 E}{(L_e / r)^2} \tag{20}$$

$$\eta = \alpha(\lambda - 0.2) \tag{21}$$

$$\alpha = \begin{cases} 0.21 & \text{"a" curve} \\ 0.34 & \text{"b" curve} \\ 0.49 & \text{"c" curve} \end{cases} \tag{22}$$

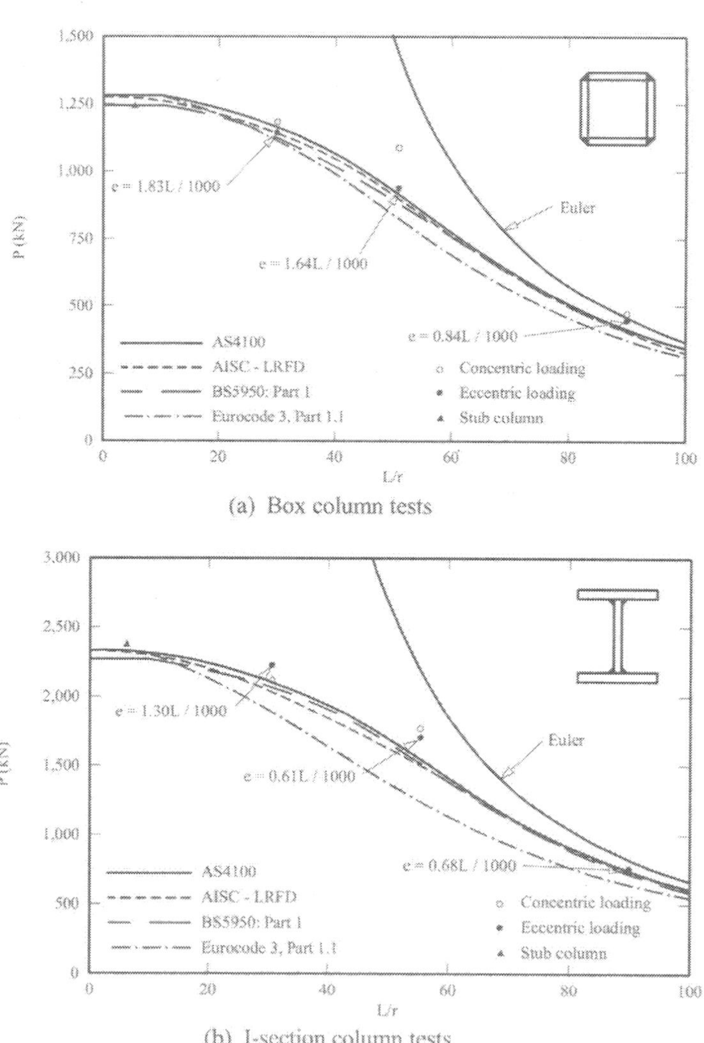

Figure 5. Test strengths and strength curves for Grade 690 MPa long columns (Rasmussen and Hancock 1995)

Equations (17-22) are implemented in Eurocode3. It is implicit in this formulation that the slenderness reduction factor $\chi(\lambda)$ is uniquely defined by the slenderness (λ).

The strength curves shown in Figs 5a and 5b were obtained on the following basis:

- In using the Australian Standard, the α_b=-0.5 curve was used in accordance with the conclusion drawn from Figs 4a and 4b. The α_b=-0.5 curve is a close fit (Rotter 1981) to the "a"-curve of the ECCS Recommendations.
- The American Specification has not adopted the multiple column curve concept, but uses a single curve which is a fit to the SSRC 2P curve (Galambos 1988). The SSRC 2P curve is based on a mean overall geometric imperfection of 1/1470 of the length (Bjorhovde 1972), and lies between the "a" and "b" curves of the ECCS Recommendations at intermediate and long column lengths but below the "b"-curve at short lengths.
- The column curves of the British Standard are defined by eqns (17-22), except that the imperfection parameter (η) is given by,

$$\eta = 0.001a\sqrt{\frac{\pi^2 E}{\sigma_y}}(\lambda - 0.2) \tag{23}$$

  rather than by eqn. (21). The constant a takes the values 2.0, 3.5 and 5.5 for the British "a", "b" and "c"-curves respectively. Consequently, in using the British Standard, the slenderness reduction factor ($\chi(\lambda,\sigma_y)$) is a function of the yield stress. The curve to be used for box and I-section (minor axis bending) columns fabricated from flame-cut plates (t<40mm) is the "b"-curve. For mild steel with a yield stress of 235 MPa this curve coincides with the ECCS "b"-curve, but for high strength steel with a yield stress of 690 MPa the imperfection parameter (eqn. (23)) becomes,

$$\eta = 0.19(\lambda - 0.2) \tag{24}$$

  and so for this value of yield stress the British "b"-curve is nearly the same as the ECCS "a"-curve, as defined by eqns (16,17), although slightly higher.
- The Eurocode3 column curves are defined by eqns (17-22). The curve specified for welded box-columns is the "b"-curve. For welded I-sections bent about their minor axis, the specified column curve is the "c"-curve. This curve is lower that those specified in AS4100 and BS5950, partly because I-sections fabricated from flame-cut plates may be designed using a higher column curve than sections fabricated from as-rolled plates according to AS4100 and BS5950, whereas no such distinction is made in Eurocode3.

In summary, the column curves of the Australian, American and British specifications to be used in the comparison with test strengths all fit closely the ECCS "a"-curve. However the column curves specified in Eurocode3 are the "b" and "c"-curves for welded box-sections and welded I-sections bent about their minor axis respectively.

As shown in Fig. 5a, the design strengths are generally in close agreement with the tests of the eccentrically loaded box columns, although the Eurocode3 design curve is conservative at intermediate and long lengths. The design strength of the Australian Standard is slightly higher than the test strength at an L/r-value of about 30. However, the loading eccentricity ($1.83L_t/1000$) used in this test was significantly higher than the nominal value of $L_t/1000$.

The Australian, American and British specifications generally agree with the tests on I-sections, although the design specifications are conservative at short to intermediate lengths, as shown in Figure 5b. The Eurocode3 design curve is significantly lower than the tests at intermediate and long lengths. This is partly because I-sections fabricated from flame-cut plates are designed using the same column curve as sections fabricated from as-rolled plates, as explained above.

The following general conclusions can be drawn from the tests results shown in Figs 5a and 5b:

- The appropriate column curve to use in the Australian standard AS4100 for welded box-columns with plate thickness less than 40mm is the α_b=-0.5 curve. This is also the appropriate curve for welded I-section columns (minor axis buckling) fabricated from flame-cut plate with thickness less than 40 mm. The α_b=-0.5 curve fits closely the "a"-curve of the ECCS Recommendations.
- If the test specimens had been fabricated from ordinary steel, the column curve would have been the α_b=0 curve according to the Australian Standard. This curve is lower than the α_b=-0.5 curve, which demonstrates that columns fabricated from high strength steel are stronger than columns fabricated from ordinary steel when compared on a nondimensional basis. The reason for this is attributed to the fact that (a) the compressive residual stresses are significantly lower than the yield stress for high strength steels compared to ordinary steel grades, and (b) when compared on a nondimensional basis and assuming negligible levels of residual stress, the strength curve for high strength steel is higher than for ordinary steel, as shown in Fig. **6**, (the parameter value n=100 represents an elastic-perfectly-plastic material).

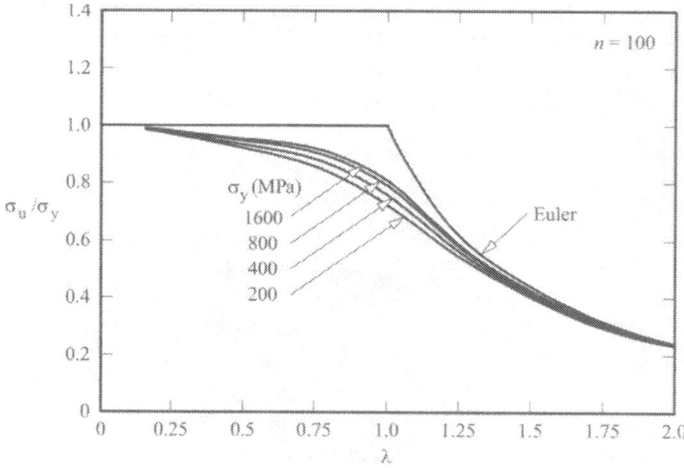

Figure 6. Effect of yield stress on non-dimensional column strength (no residual stress)

- The design curves of the Australian, American and British specifications are nearly identical for welded high strength box-columns and I-sections bent about their minor axis, and are shown to be in close agreement with the test strengths. However, the Eurocode3 design curves are conservative compared with the tests. This is because

Eurocode3 uses the ECCS "b"-curve for welded box-sections, whereas curves similar to the higher ECCS "a"-curve are specified in the Australian, American and British specifications. The British curve closely fits the ECCS "a"-curve, which is partly because the imperfection parameter is a function of the yield stress in the British Standard. For welded I-sections bent about their minor axis, the Eurocode3 design curve is conservative because it was based on the ECCS "c"-curve whereas the Australian, American and British specifications are based on curves similar to the ECCS "a"-curve. In the comparison, the Eurocode3 design curves were obtained from Section 5.5.1 of Eurocode3. Annex D of Eurocode3 allows 460 MPa yield hot-rolled I-sections bent about their minor axis to be designed to the ECCS "a"-curve which is consistent with the Australian, American and British specifications. The test results shown in Figs 5a and 5b suggest that the Eurocode3 "a"-curve can also be used for welded I-sections bent about their minor axis and welded box sections.

6.2.5 Limitations on the Use of Quenched and Tempered Steels in Structural Design

The Australian standard AS4100 (1998) limits the yield stress to 450 MPa. It does not rule out the use of high strength quenched and tempered steels but requires the yield stress be taken as 450 MPa for steels with nominal yield stress values greater than 450 MPa. The research described in Rasmussen and Hancock (1992; 1995) was undertaken in an effort to allow structural design of high strength steels with yield stress values up to 690 MPa to be covered by AS4100. However, such amendment has not yet been made.

The British standard BS5950, Part 1, (BSI 2001) includes steel grades with yield stress values up to 460 MPa. This is consistent with Annex D of Eurocode3, Part 1.1, which allows design of quenched and tempered steels with yield stress values of 420 MPa and 460 MPa. However, according to Section 3.2.2.2 of Eurocode3, Part 1.1, the use of plastic analysis requires that $\sigma_u/\sigma_y \geq 1.2$, where σ_u is the ultimate tensile strength, $\varepsilon_t \geq 15\%$, where ε_t is the total strain at failure, and $\varepsilon_u \geq 20\varepsilon_y$ where ε_u and ε_y are the strains corresponding to the ultimate tensile strength and yield stress respectively. Some of the 460 grade steels included in Annex D do not satisfy these requirements.

The main AISC-LRFD Specification (1999) allows quenched and tempered steels to be used in design, including steels to ASTM A852 (2001)with a yield stress of 485 MPa and ASTM A514 (2000) with a yield stress of 690 MPa. However, these steels are not permitted in the AISC Seismic Provisions for Structural Steel Buildings (AISC 1997). Furthermore, Section 5.1 of the main specification does not permit plastic design for steels with yield stress greater than 448 MPa. This restriction was imposed because of lack of information on the moment rotation behaviour of quenched and tempered steel beams. Early tests by McDermott (1969) showed that Grade 690 MPa steel beams to ASTM A514 may have limited rotation capacity.

6.3 Cold-reduced steels

6.3.1 General

The steels here referred to are light-gauge steels typically to AS1397 (1993), ASTM A653 (2001), ASTM A792 (2002) and ASTM A875 (2001), EN10214 (1995), EN10215 (1995) and EN10292 (2000). The steels included in these specifications range in yield stress from about 200 MPa to 550 MPa, depending mainly on the amount of cold-work induced in rolling the steel and subsequent heat treatment. The thickness range varies from about 0.4 mm up to about 5 mm with the highest strengths obtained for the thinner gauges. The enhanced strength is derived from strain-hardening induced by cold-reducing the thickness.

Cold-reduced high strength steels are used in a wide range of applications. For instance, 1-3 mm thick material is commonly used for purlins in roofing and uprights in steel storage racking among many other applications, while the thinner gauges with thickness less than 1 mm are used frequently for floor decking, as well as roof and wall sheeting. In Australia, cold-reduced steels with thickness less than 1 mm are now also being used for roof trusses, and residential steel framing.

The maximum yield stress considered in the European material specifications EN10214, EN10215 and EN10292 is 400 MPa, while high strength steels with a yield stress value of 550 MPa are included in the Australian specification AS1397 and the American ASTM Standards A653, A792 and A875. Common to Eurocode3, Part 1.3, (Eurocode3-1.3 1996), the American Iron and Steel Institute Specification (AISI 1997) and the Australian specification for cold-formed steel structures (AS/NZS4600 1996) is that structural steels are required to have sufficient ductility to provide safe and serviceable structures. Section 3.2.1 of Eurocode3, Part 1.3, requires that the ratio (σ_u/σ_y) of ultimate tensile strength to yield stress is greater than or equal to 1.1. In Section A3.3 of the AISI Specification and Section 1.5.1.5 of AS/NZS4600, this requirement is relaxed to 1.08. However, further requirements are imposed on the elongation (total, uniform and/or local), as obtained from tensile tests. If these ductility requirements are met, the design strength provisions of the three specifications can be used without restriction. This applies to the steels included in the abovementioned materials specifications for all yield stress values up to 500 MPa. The design of these relatively ductile steels is covered elsewhere in this lecture series.

Of particular interest is the design of the Australian G550 steel to AS1397, which has a nominal yield stress of 550 MPa and a thickness range from 0.42 mm to 1 mm. This steel corresponds closely to the ASTM A653, A792 and A875 Grade 80 steels. The σ_u/σ_y-ratio for these steels is generally less than 1.08, depending on the thickness. Hence, specific design guidelines are included for these steels in the AISI Specification and AS/NZS4600, as described briefly in Section 6.3.4.

6.3.2 The Cold-reducing and Galvanising Process

The following excerpt of the fabrication process of high strength cold-reduced steels is based on Rogers and Hancock (1997).

Initially the sheet steel is rolled to size in a hot strip mill with finishing and coiling temperatures of approximately 940°C and 670°C respectively. The hot-worked coil, with typical yield stress of 300 MPa, is later uncoiled, cleaned, trimmed to size and fed through a cold-reducing mill. High compressive roller forces and strip tension gradually reduce the thickness of the steel sheet to the desired dimension. For 0.42 mm and 0.6 mm thick G550 steel, the thickness is reduced by 75 to 85%.

The milling process causes the grain structure of cold-reduced steels to elongate in the rolling direction, which produces an increase in material strength and a decrease in ductility. The grain distortion increases with cold-reduction, however, it is possible to change the distorted grain structure and control the steel properties through subsequent heat treatment. The temperature may be increased beyond the recrystallisation temperature in which case the grain structure returns to its original state as in the case of G300 steel, or may be stress relief annealed as in the case of G550 steel. Stress relief annealing involves heating the steel to below the recrystallisation temperature, holding the steel to ensure uniform temperature through the thickness and then slowly cooling the steel. Annealing is carried out in a hot dip coating line prior to application of either a zinc or aluminium/zinc coating. Upon final cooling, the steel sheet proceeds through a tension leveling or skin-passing mill to improve the finish quality and flatness of the coil. The mechanical properties of the finished coil depend on the amount of cold-reduction, stress relief annealing temperature and finishing treatment (tension leveling or skin-passing).

6.3.3 The Stress-strain Curve

The strain-hardening capacity and ductility reduces with the amount of cold-working. For the thinnest steels of about 0.4 mm thickness, the ductility is of the order of a few percent, and there is no difference between the yield stress and the ultimate tensile strength. Table 7 shows the nominal values of yield stress, tensile strength and minimum elongation of Australian cold-reduced steels to AS1397 (1993).

Table 7. Yield stress, ultimate tensile strength and elongation for longitudinal tensile tests according to AS1397 (1993)

Steel grade designation#	σ_y MPa	σ_u MPa	Min. elongation* (%) L_o=50 mm	Min. elongation* (%) L_o=80 mm
G250	250	320	25	22
G300	300	340	20	18
G350	350	420	15	14
G450	450	480	10	9
G500	500	520	8	7
G550	550	550	2	2

The letter "G" indicates that the mechanical properties have been achieved or modified by in-line heat treatment prior to application of either a zinc or aluminium/zinc coating. The three digits following are the nominal yield stress in MPa

* Applies to test pieces equal to or greater than 0.6 mm thick. L_o=original gauge length

Figure 5 shows typical stress-strain curves for G550 steel to AS1397 with nominal thickness varying from 0.95 mm to 0.42 mm. Several observations can be made from these figures:

- The stress-strain curves do not have upper and lower yield values as for mild steel.

- The stress-strain curves are rounded with a proportionality stress of about 70% of the yield value. For this reason, the yield stress is often determined as the 0.2 % proof stress. The proportionality limit depends on the amount of cold reduction, stress relief annealing and the finishing process (tension leveling or skin passing).
- The ductility decreases with decreasing thickness: The $\varepsilon_u/\varepsilon_y$-ratio varies from about 13 for 0.95 mm thick sheet to about 1 for a 0.42 mm thick sheet.
- For the 0.6 m and 0.42 mm thick sheets, the strain-hardening capacity is eroded and the yield stress equals the ultimate tensile strength.

The mechanical properties of cold-reduced steels are described in further detail in Chapter 2 of Hancock et al. (2001)

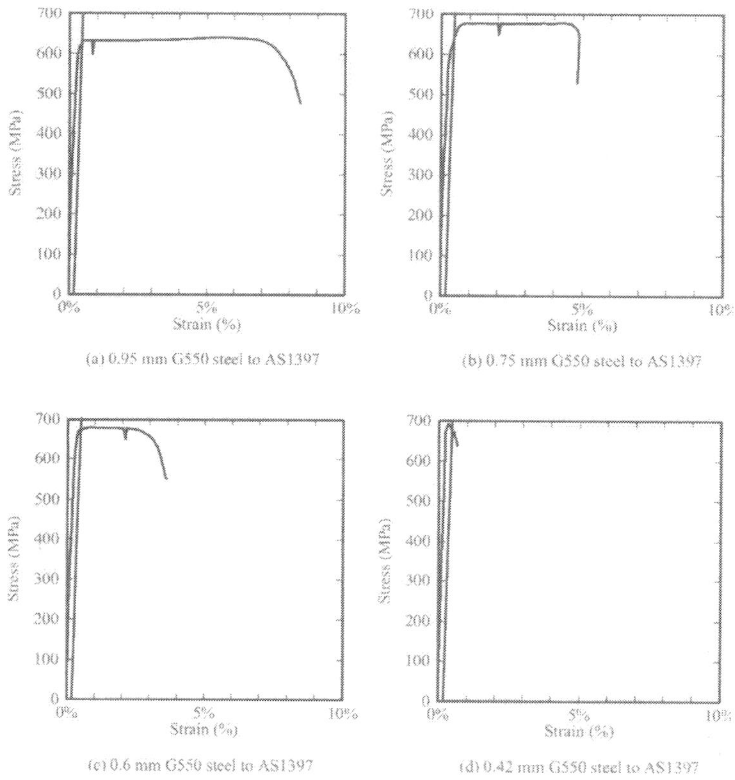

(a) 0.95 mm G550 steel to AS1397 (b) 0.75 mm G550 steel to AS1397

(c) 0.6 mm G550 steel to AS1397 (d) 0.42 mm G550 steel to AS1397

Figure 8. Stress-strain curves of G550 steels varying in thickness from 0.95 mm to 0.42 mm.

6.3.4 Structural Design of Cold-reduced 550 MPa Steels

Because of the limited ductility and lack of research data to support the structural use of thin gauge cold-reduced steels, current Australian design provisions (AS/NZS4600 1996) impose limits on the nominal yield stress and ultimate tensile strength of G550 steel to be used in structural design. The American specification (AISI 1997) goes further and rules out the use of thin-gauge cold-reduced Grade80 steels for main structural members, but allows their use in roof and wall sheeting as well as floor decking.

Section 1.5.1.5(b) of the Australian cold-formed steel structures standard AS/NZS4600 (1996) requires the yield stress and ultimate tensile strength of G550 steel less than 0.9 mm thick be reduced to 75% of the nominal values. A similar requirement is made in the AISI Specification (AISI 1997) and its Supplement (AISI 1999) for Grade80 steels to ASTM A653, A792 and A875. A less stringent requirement is made in the Supplement, where the reduction factor varies between 0.75 and 1 depending on the slenderness of the panel. However, the AISI Specification limits the application of Grade80 steels to multiple-web configurations used in roofing, siding and floor decking.

Research is ongoing into the structural use of cold-reduced steels Grade 550 steels. The research so far indicates that the limited ductility of G550 steel does not by itself lead to lower strength. However, because of the thinness of the steel, new failure modes are observed which require modifications to existing design provisions for connections (Rogers and Hancock 1998) and compression members (Yang and Hancock 2001).

References

AISC (1997). Seismic Provisions for Structural Steel Buildings, American Institute of Steel Construction, Chicago.

AISC-LRFD (1999). Specification for Structural Steel Buildings, Load and Resistance Factor Design, American Institute of Steel Construction, Chicago.

AISI (1997). Load and Resistance Factor Design Specification for Cold-formed Steel Structural Members, American Iron and Steel Institute, Washington, DC.

AISI (1999). Supplement No. 1 to the Specification for the Design of Cold-formed Steel Structural Members, American Iron and Steel Institute, Washington, DC.

AS1397 (1993). Steel sheet and strip—Hot-dipped zinc-coated or aluminium/zinc-coated, Standards Australia, Sydney.

AS4100 (1998). Steel Structures, AS4100, Standards Australia, Sydney.

AS/NZS4600 (1996). Cold-formed Structures, AS/NZS 4600, Standards Australia, Sydney.

ASTM-A514 (2000). Standard Specification for High-Yield-Strength, Quenched and Tempered Alloy Steel Plate, Suitable for Welding, A514, American Society for Testing and Materials, Philadelphia.

ASTM-A653 (2001). Standard Specification for Steel Sheet, Zinc-Coated (Galvanized) or Zinc-Iron Alloy-Coated (Galvannealed) by the Hot-Dip Process, A653, American Society for Testing and Materials, Philadelphia.

ASTM-A792 (2002). Standard Specification for Steel Sheet, 55 % Aluminum-Zinc Alloy-Coated by the Hot-Dip Process, A792, American Society for Testing and Materials, Philadelphia.

ASTM-A852 (2001). Standard Specification for Quenched and Tempered Low-Alloy Structural Steel Plate with 70 ksi [485 MPa] Minimum Yield Strength to 4 in. [100 mm] Thick, A852, American Society for Testing and Materials, Philadelphia.

ASTM-A875 (2001). Standard Specification for Steel Sheet, Zinc-5% Aluminum Alloy-Coated by the Hot-Dip Process, A875, American Society for Testing and Materials, Philadelphia.

Bjorhovde, R (1972). Deterministic and Probabilistic Approaches to the Strength of Steel Columns. Bethelehem, PA, Lehigh University.

BSI (2001). Structural use of Steelwork in Building. Specification for Materials, Fabrication and Erection: Hot-Rolled Sections, British Standards Institution, London.

ECCS (1977). *European Recommendations for Steel Construction*, Proceedings of European Convention for Constructional Steelwork (ECCS), Brussels.

EN-10113 (1997). Hot-rolled products in Weldable Fine Grain Structural Steels, EN-10113, European Committee for Standardisation, Brussels.

EN-10214 (1995). Continuously hot-dip zinc-aluminium (ZA) coated steel strip and sheet, EN-10214, European Committee for Standardisation, Brussels.

EN-10215 (1995). Continuously hot-dip zinc-aluminium (AZ) coated steel strip and sheet, EN-10215, European Committee for Standardisation, Brussels.

EN-10292 (2000). Continuously hot-dip coated strip and sheet of steels with higher yield strength for cold forming, EN-10292, European Committee for Standardisation, Brussels.

Eurocode3-1.1 (1992). Eurocode3: Design of Steel Structures, Part 1.1: General Rules and Rules for Buildings, ENV-1993-1-1, European Committee for Standardisation, Brussels.

Eurocode3-1.3 (1996). Eurocode3: Design of Steel Structures, Part 1.3: Cold Formed Thin Gauge members and Sheeting, ENV-93-1-3, European Committee for Standardisation, Brussels.

Fukumoto, Y and Itoh, Y (1984). "Basic Compressive Strength of Steel Plates from Test Data." *Japanese Society of Civil Engineers, Transactions* **No. 344**: 129-139.

Galambos, T (1988). *Guide to Stability Design Criteria for Metal Structures*. New York, John Wiley and Sons.

Hancock, GJ, Murray, TM and Ellifrit, DS (2001). *Cold-formed Steel Structures to the AISI Specification*. New York, Marcel Dekker, Inc.

McDermott, J (1969). "Plastic Bending of A514 Steel Beams." *Journal of the Structural Division, American Society of Civil Engineers* **95**(ST9): 1851-1871.

Moxham, K (1971). Buckling Tests on Individual Welded Steel Plates in Compression, *Technical Report NO CUED/C-Struct/TR3*, Department of Civil Engineering, Cambridge University, Cambridge.

Murphy, W (1964). Mettallurgical Advantages of Heat-treated Steels, *Design and Engineering Seminar - 1964, Publication No. ADUSS91-1008*, US Steel Corporation.

Nishino, F and Tall, L (1970). Experimental Investigation of the Strength of T-1 Steel Columns, *Fritz Engineering Laboratory Report No 290.9*, Lehigh University, Bethelehem, PA.

Nishino, F, Ueda, Y and Tall, L (1966). Experimental Investigation of the Buckling of Plates with Residual Stress, *Fritz Engineering Laboratory Report No 290.3*, Lehigh University, Bethelehem, PA.

Rasmussen, K and Hancock, G (1992). "Plate Slenderness Limits for High Strength Steel Sections." *Journal for Constructional Steel Research* **54**(1-3): 73-96.

Rasmussen, K and Hancock, G (1995). "Tests of High Strength Steel Columns." *Journal for Constructional Steel Research* **34**(1): 27-52.

Rogers, C and Hancock, G (1997). Bolted Connection Tests of Thin G550 and G300 Sheet Steels, *Research Report No. R749*, Department of Civil Engineering, University of Sydney, Sydney.

Rogers, C and Hancock, G (1998). "Bolted Connection Tests of Thin G550 and G300 sheet steels." *Journal of Structural Engineering, American Society of Civil Engineers* **124**(7): 798-808.

Rondal, J and Maquoi, R (1979). "Single Equation for SSRC Column-strength Curves." *Journal of the Structural Division, American Society of Civil Engineers* **105**(ST1): 247-250.

Rotter, J (1981). *A Simple Approach to Multiple Column Curves*, Proceedings of Metal Structures Conference, Newcastle, Australia.

Usami, T and Fukumoto, Y (1982). "Local and Overall Buckling of Welded Box Columns." *Journal of the Structural Division, American Society of Civil Engineers* **108**(ST3): 525-542.

Yang, D and Hancock, G (2001). *Compression Tests of Box Shaped Cold-Reduced High Strength Steel Sections*, Proceedings of Sixth Pacific Structural Steel Conference, Beijing.

Chapter 7: Residential Buildings

J. M. Davies

The Manchester School of Engineering, University of Manchester, Manchester, England
E-mail: jmdavies@fs1.eng.man.ac.uk

Residential and Commercial Buildings

7.1 Light gauge steel framing systems for residential and commercial buildings

7.1.1 Introduction

All over the industrialised world, the use of light gauge steel framing for residential and other low-rise construction is increasing rapidly. In some countries, such as the UK and most of Europe, this is from a small base. In others, such as Australia, Japan, the USA and Canada, there is a well-developed industry with well-established practices. This is encouraging news for those of us in Europe who are interested in light gauge steel construction.

The present generation of steel framed house construction systems has evolved from traditional construction to timber frame to steel frame without any significant change in either the architecture or the interior and exterior finishing materials. The early timber framed designs simply involved placing a framing system based on 4 inch × 2 inch vertical timber members at 18 inch centres behind traditional facades. In many current steel frame designs, the timber has merely become 100 mm × 50 mm cold formed steel channel sections (wall studs) at 450 mm (or possibly 600 mm) centres. This is disappointing because a facade designed for construction in (say) load-bearing brickwork is unlikely to be optimal for construction based on steel framing.

However, this is changing and indeed it has to change. Traditional construction designed to conceal a steel frame is not the way forward into the 21st Century. This series of papers attempts to consider what the structural solutions behind the architecture will look like. It is implicit in this discussion that the steel structures will not remain concealed for ever. The time will surely come when house owners will be proud to own their steel framing.

7.1.2 Primary features of modern steel framed construction

There are a number of features that make steel framed construction essentially different from traditional construction. Different authors may place these in a different order of priority. However, the list which follows is what the author of this paper considers should be the distinctive features.

It is clear that the primary feature ought to be <u>factory-based prefabrication</u>. In every other walk of life, (e.g. cars, TV sets, computers) the trend has been to dramatically reduce the number of components in the final installation. The construction industry has been surprisingly slow to follow this trend. However, sooner rather than later, this has to happen with house and related low-rise construction too. The economic driving forces for this change are speed of construction linked to levels of quality control that are only possible under factory conditions. Another significant factor is the increasing shortage of skilled tradesmen on site.

The next important parameter has to be dry construction. The dirt, dust, shrinkage and general lack of precision associated with hand-laid masonry, in-situ concrete and wet plaster should be unacceptable in modern construction.

The architect of the future will have to accept a stricter modular discipline. This will primarily affect the dimensions and locations of openings such as doors and windows. Rational steel frame designs will only be obtained when the engineer works side by side with the architect to realise a modular frame design with members on a regular grid and sensible paths to ground for both vertical and wind loads.

This requirement leads naturally to the related requirement for simplified engineering design. The engineering design of a complete light gauge steel framed house from basic code of practice principles is unacceptably labour intensive. This will require that more attention is given to prescriptive methods of design so that, ultimately, a steel framed solution will require no more engineering input than the timber-framed or load bearing masonry competition.

Simpler engineering has logically to be associated with simpler detailing of connections. In particular, this entails taking a long look at how wind forces are transmitted through the structure to the foundations. Many of the rather "fussy" details which tend to characterise current designs are concerned with this aspect of design whereas the vertical load bearing structure is relatively simple.

Ideally, the steel frame should be fully integrated with both the external and internal wall panel systems as well as the floor systems so that they are mutually supporting. This leads naturally to considerations of stressed skin design. A special case of this, which is considered in more detail later, is to use a steel cassette wall system with the wide flange outwards in "cold frame" construction. The steel cassette is thus both the frame and the outer cladding combined and just needs a suitable internal finish to complete the wall construction.

Finally, but by no means least in economic importance, the construction should provide a usable roof space. This is not logically essential. However, the point is that a habitable roof area can be provided at little additional cost. When this is done, it can be the single factor that makes steel framed construction significantly cheaper than the alternatives.

7.1.3 Framing systems for house construction (CSSBI, 1987; Grubb and Lawson, 1997)

Figure 1 shows typical steel framed construction (CSSBI, 1987)

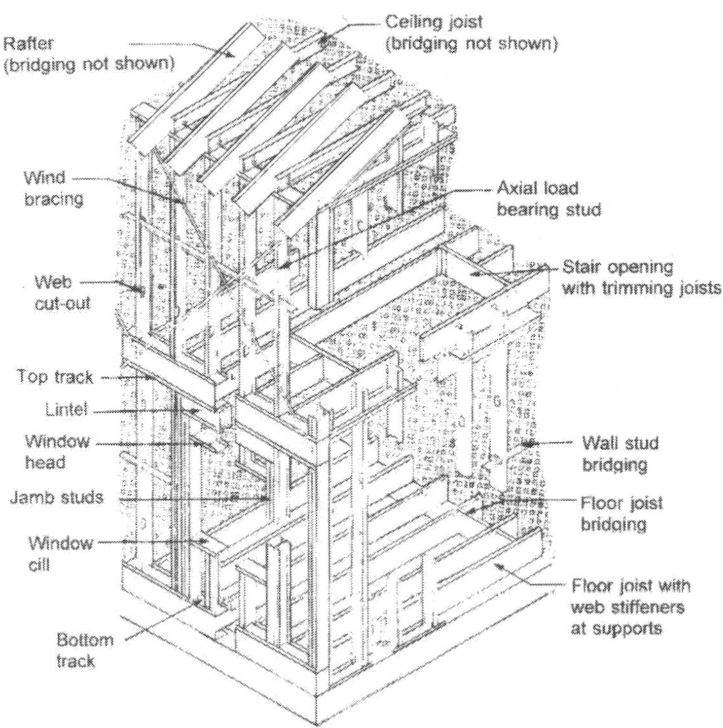

Figure 1. Typical light gauge steel framing for house construction

The basic elements of construction are:

1. Walls: Typically 100 × 50 × 1.5 cold-formed steel studs with top and bottom tracks, framing for window and door openings, etc.
2. Floors: Typically with 150 × 50 × 2.0 cold-formed steel joists
3. Roof: Various systems including cold-formed steel rafters as shown.
4. Wind bracing:

There are two basic framing systems in current use which are usually defined as "platform" and "balloon". Additional fundamental differences arise as a consequence of the degree of prefabrication that is used and whether the primary fixing technology is welding or the use of mechanical connections (eg bolts or screws). Differences may also arise in the way that in plane wind shear (racking) forces are resisted. Most of the options can be found in current practice.

The main framing systems referred to above are generally constructed using vertically spanning 'C' shaped cold-formed sections, which are generally referred to as wall studs. A completely different system, which will be described later, uses cassette sections. With cassette sections, the two basic platform and balloon framing systems remain available.

In platform construction, as shown in Figure 2(a), the structure is built storey by storey so that each floor can serve as a working platform for the construction of the floor above. The walls are not structurally continuous and loads from the walls above are generally transferred through the floor structure to the walls below. The wall studs are connected to horizontal tracks top and bottom and the floor joists are seated on the top track of the studs below. Sufficient stiffening is incorporated in the connection to ensure the safe transfer of vertical load through the floor construction.

In balloon construction, as shown in Figure 2(b), the wall panels are continuous over two or more storeys and the floors are attached to them. It follows that loads from the floors above pass down the load-bearing studs without affecting the incoming floor construction.

Both of the above methods of framing may use 'stick' construction, in which individual members are assembled on site, or they may use varying degrees of prefabrication. The advantages of stick construction are (Grubb and Lawson, 1997):

- The system can accommodate larger construction tolerances
- The workshop facilities associated with modular construction are not necessary
- Simple construction techniques without heavy lifting equipment may be used
- Members can be densely packed for transport

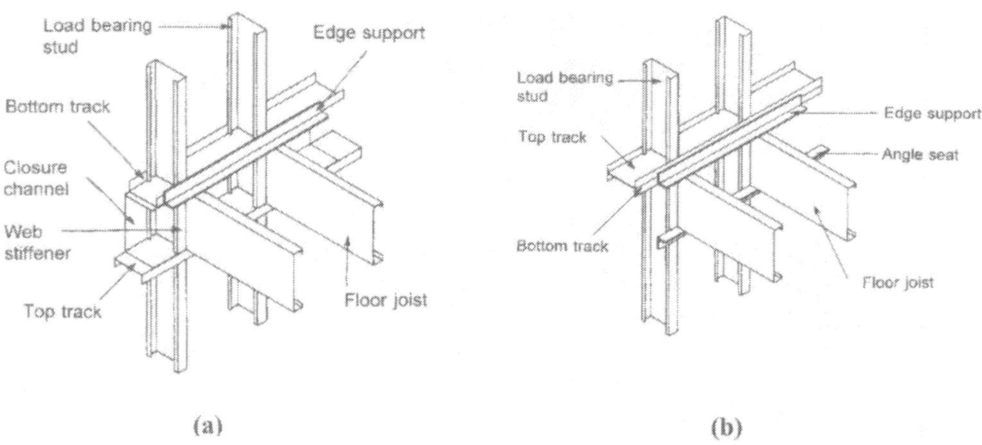

Figure 2: Framing details with (a) platform and (b) balloon construction (CSSBI 1987)

Members may be cut to length on site with a hand-held saw and connections may be made by any of the usual mechanical methods such as bolts, self-drilling, self-tapping screws or blind rivets.

Alternatively, in 'panel' construction, wall and floor sub-frames and roof trusses may be prefabricated in the factory in purpose-made jigs. Factory-made connections may be welded or they may employ any of the conventional mechanical methods such as bolts, screws or rivets. Welding has the advantage of simplicity and permanence but has the disadvantage of destroying the advantage of pre-galvanising at critical locations in the structure. An attractive modern method of making permanent connections without the disadvantages associated with welding is press-joining. This is considered later. Thermal insulation and some of the lining and finishing materials may also be applied to the steel sub-frames in the factory. Factory-prefabricated units are then generally connected together on site by bolting.

The main advantages of panel or sub-frame construction are (Grubb and Lawson, 1997):

- Speed of erection
- Factory standards of quality control during fabrication of the units
- Reduction of site labour costs
- Scope for automation in factory production

Prefabrication of wall, floor and roof units is taken a stage further in modular or 'volumetric' construction. Here, light gauge steel boxes, which may for example be hotel room units, are completely prefabricated in the factory before being delivered to the site. It is usual for all internal finishes, fixtures and fittings, and even the carpets, curtains and furniture, to be fitted in the factory. On site, the units are stacked side-by-side and up to several storeys high on prepared foundations and service connections made to form the complete structure. Nowadays, "many out-of-town" hotels and motels are built in this way.

7.1.4 Cassette wall construction

The cassette wall system has been pioneered in France by the company 'Produits Acier Batiment' (PAB) under the name 'CIBBAP'. A fuller description of this form of construction with some case studies will be given later.

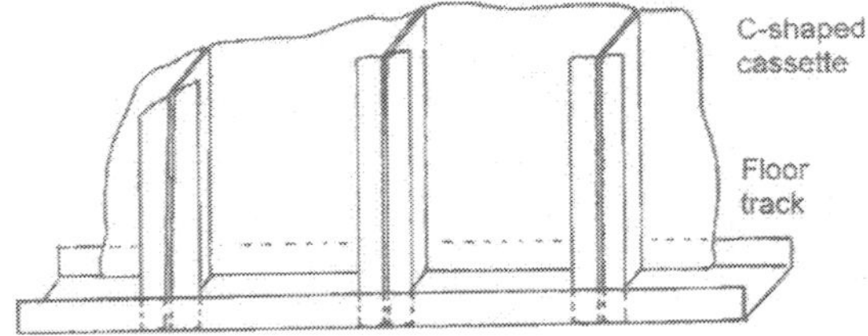

Figure 3. Cassette wall construction

The basic arrangement for cassette wall construction is shown in Fig. 3. Lipped C-shaped cassettes span vertically between top and bottom tracks. Either platform or balloon construction may be used although practical examples to date have generally used balloon construction. Cassette construction may be viewed as stud construction integrally combined with a metal lining sheet to provide a metal frame together with a flat steel wall. Light gauge steel cassettes do not readily lend themselves to stick construction and it is usual to prefabricate complete sub-frames comprising the cassettes and their top and bottom tracks in the factory.

7.1.5 Floor construction

Floor construction is logically lightweight and dry. However, there are circumstances where heavier construction is specified, usually to meet requirements for fire and/or sound insulation.

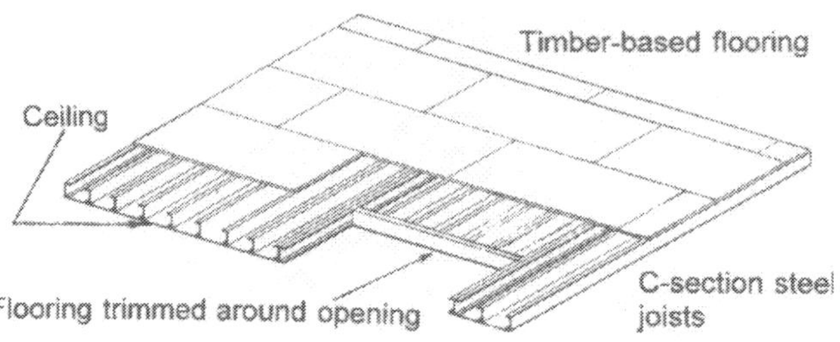

Figure 4. Lightweight steel floor construction

There are, therefore, three generic systems for steel-based floor construction:

- Steel joists (usually C- or Z-section) with a timber-based deck as shown in Fig. 4. The joists are usually on a module which coincides with studs on the supporting elevations.
- Trapezoidally profiled steel deck supported on steel primary beams and carrying a timber-based walking surface.
- Composite (steel-concrete) deck supported on steel primary beams.

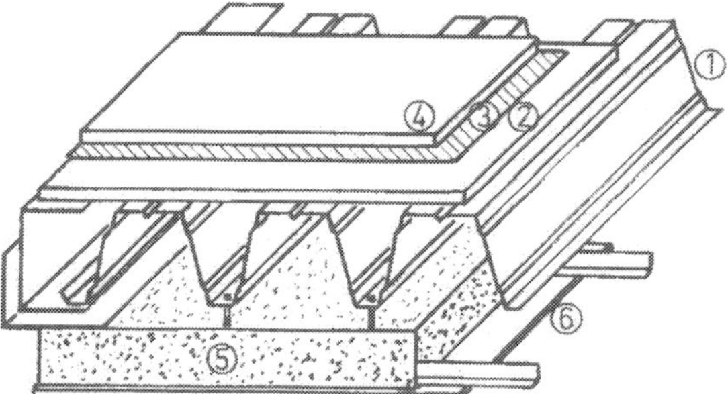

1. Steel deck
2. Transfer panel
3. Resilient layer
4. Walking surface (floor covering over)
5. Insulation
6. False ceiling

Figure 5. Typical internal structure of a dry floor on a steel deck (Engel, 1998)

Superficially, the structural engineering design is conventional and elementary. However, joist designs are rarely based on the bending resistance of the cross-section. Normally, the serviceability criteria of deflection and vibration control the design. Suspended floors can sometimes resonate with the vibrations induced by human footsteps and, although this does not usually lead to failure, it can be a source of significant discomfort. This problem tends to become more acute when lightweight construction is used.

Engel (1998) discusses the advantages of dry floor construction and shows how the requirements of building physics may be achieved. He offers the scheme shown in Fig. 5 as a tried and tested means of providing a performance which meets all of the criteria of strength, stiffness, serviceability (vibration), acoustic and thermal insulation. Grubb and Lawson (1997) also give examples of internal floors of dry construction with varying degrees of sound insulation.

7.1.6 Roof construction

A steel truss with similar proportions to a conventional gang-nailed timber truss is often about six times as strong as the timber equivalent and is uneconomic at modest spans and load levels. For this reason, when a habitable roof space is not required, steel framed houses are often completed with timber roof trusses. Steel trusses can, however, be used economically when they are spaced at wider centres with purlins spanning between the trusses. All-steel solutions become much more favourable when a usable roof space is required. Figure 6 (Grubb and Lawson, 1997) shows a typical solution in which attic frames are spaced at 600 mm centres and battened and tiled in a conventional manner. An alternative system could use sandwich panels either spanning from eaves to ridge or supported on intermediate purlins.

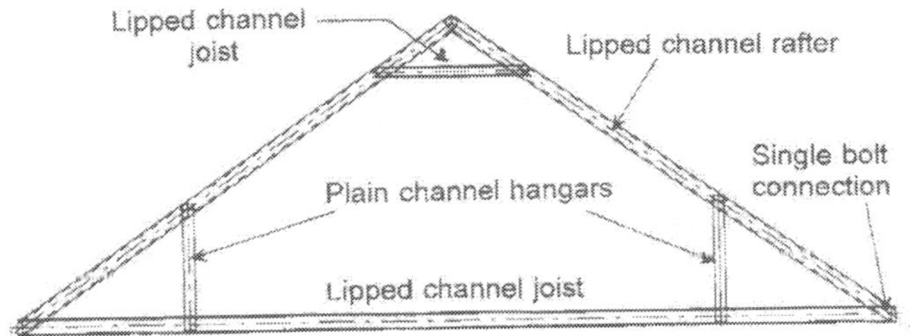

Figure 6. Steel roof construction with usable loft space

7.1.7 Structural stability

One of the fundamental requirements in light steel framed construction is the provision of resistance to the in-plane forces arising as a result of horizontal (wind) loads on the building. There are two components of this problem, as shown in Figure 7:

- The floor and roof planes must be able to act as diaphragms in plan to transmit wind forces on the windward and leeward elevations back to the walls parallel to the direction of the wind.
- The walls themselves must be able to transmit the in-plane forces from the floor and roof planes down to the foundations. It is then found that significant uplift forces may arise at the leeward end of the wind-frames. Providing an adequate tie to adequate foundations is another aspect of this problem.

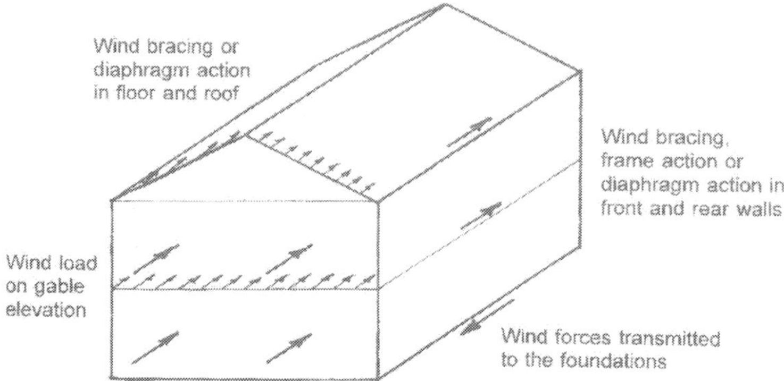

Figure 7. Resistance to horizontal wind loads in house construction

Both of these requirements can be problematical and the critical importance of diaphragm action in the first floor should be particularly noticed. Most structural systems place a high reliance on this element.

Floor construction often takes the form of a timber-based product supported by steel joists. It is sometimes assumed that this form of construction can provide the necessary diaphragm action without formal calculations. This is a highly dangerous assumption! Diaphragm action in the floor is fundamental to the stability of the structure and the in-plane forces can be very significant. The situation is not helped by the presence of a stair well. The diaphragm action of the floor must always be properly calculated when it will be found that the main problem is in providing the required shear connection to the primary structure at the same time bearing in mind the absence of any 'seam strength' between the individual sheets of floor material.

With light steel framing, a more reliable solution is to provide a light gauge steel floor deck. This allows the formal calculation procedures of stressed skin design to be used. When this calculation procedure is used, it is found that the required shear strength can be readily achieved, helped by the provision of seam fasteners. The demands made on the relatively reliable metal to metal fasteners in this form of construction highlight the potential difficulties with timber floors.

It is necessary to make similar provision at the roof level and, here, designers have tended to be more circumspect and often provide a wind girder as shown in Fig. 8. However, if, as is readily possible with light gauge steel framing, it is intended to utilise the roof space, an alternative and advantageous solution is to take advantage of the diaphragm action of the floor deck. The considerations then become similar to those discussed above.

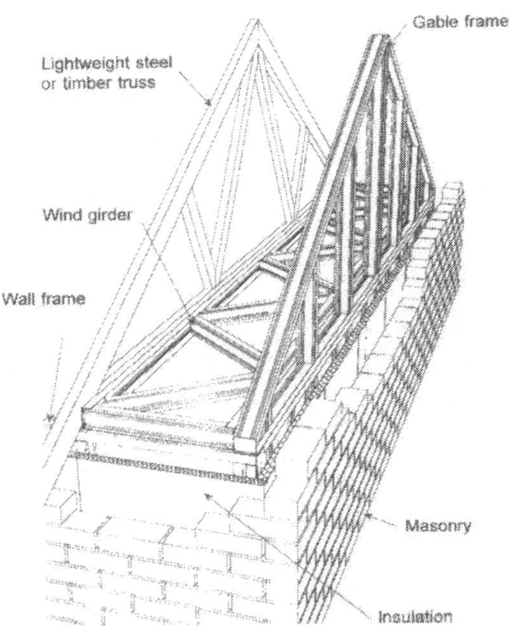

Figure 8. Wind girder at eaves level (Grubb and Lawson, 1997)

There are several methods available for providing in-plane stability in vertical walls:

- The most common method is the use of X-bracing whereby crossed flat straps of thin steel pass over the faces of the studs, as shown in Fig. 9(a). These straps act only in tension so that only one is active at any given time. They are usually nominally fixed to the studs in order to reduce their tendency to sag. They must, of course, be properly anchored at their ends in order to transfer the calculated tensile force to the primary structure.

- An alternative system is K-bracing which takes the form of C-sections fixed within the depth of the stud walls as shown in Fig. 9(b). These members act in either tension or compression and, together with the adjacent studs, form a vertical truss. They again require appropriate connections at their ends. The additional compressive stresses in the leeward member of this truss may require that the adjacent studs have a larger section than the norm. In the example shown, where the wind trusses are relatively narrow, providing the resistance to the uplift forces in the leeward vertical member may be problematic.

  Fig 9(b), taken from Grubb and Lawson (1997), provides a nice illustration of a fundamental point. The designers of the steel framing have evidently been provided with an existing elevation, presumably designed for masonry construction, and instructed to fit a steel frame into it. The result can hardly be described as satisfactory. There is little evidence of the modular coordination necessary for economic steel framed construction and the somewhat random arrangement of windows and doors in the right hand half of the diagram makes the provision of sensible load paths more difficult. In particular, little thought has been given to the provision of resistance to in plane wind shears and arrangement of the two narrow vertical wind trusses is far from ideal. The contrast with Fig. 9(a) is clear.

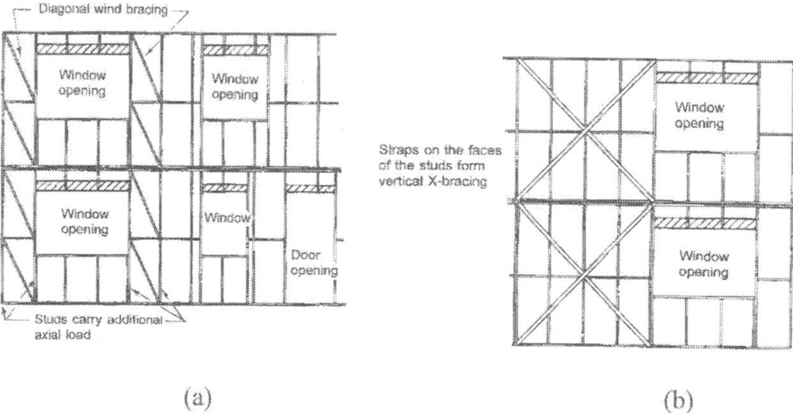

Figure 9. Elevations of a building with (a) vertical 'X' bracing and (b) vertical 'K' bracing

As a practical point, it is important to detail either form of diagonal bracing system to avoid significant eccentricities at the connections. Whichever of these first two systems is adopted, it is clear that designing and detailing wind bracing in the plane of the walls is a significant expense and disadvantage. Evidently, X-bracing should be used wherever possible and current practice is to use this method for the party walls of semi-detached and terraced houses and for other gable walls where there are few door and window openings. However, the currently popular building style results in front and rear elevations which contain much more opening than solid wall. In such cases, X-bracing of the type shown in Figure 9(a) may not be possible and it is necessary to seek alternative solutions.

- The wall structure can act as a 'rigid-jointed' plane frame. When bolted stick construction is used for site assembly, this is probably essential for elevations with significant openings. It is difficult to see how vertical wind trusses, as shown in Fig. 9(b), could be fabricated satisfactorily on site. Furthermore, consider the elevation shown in Fig. 10, which is not a figment of the author's imagination but the elevation of an actual steel-framed house. How else could in-plane strength be obtained in such a case?

It is important to appreciate here that bolted joints in light-gauge steel construction are by no means rigid to the extent that a convention rigid-jointed plane frame analysis is unlikely to be valid. It is essential to estimate the flexibility of the connections and to include these in a semi-rigid analysis. Our understanding of semi-rigid construction has improved dramatically in recent years so that this requirement need present no significant practical difficulties.

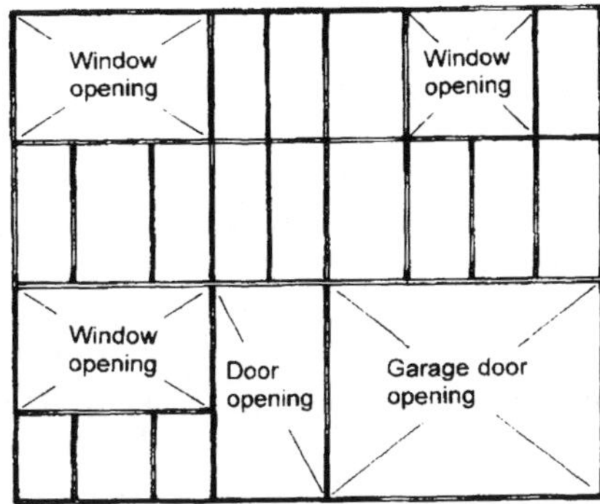

Figure 10. Front elevation with large openings

Elevations such as are shown in Figs. 9(b) and 10 are quite typical of current practice and clearly present difficulties in providing practical and economic steel framed solutions in the presence of wind shear. There is an alternative solution to this problem. If the floor and roof (or more advantageously the loft floor) are considered to be 'cantilever diaphragms', the wind shear in the problematic front elevation can be transferred to the other three elevations where there may be fewer openings. The static principles of this solution are shown in Fig. 11. Clearly, if this approach is adopted, it is necessary to pay particular attention to the diaphragm strength of the floor construction.

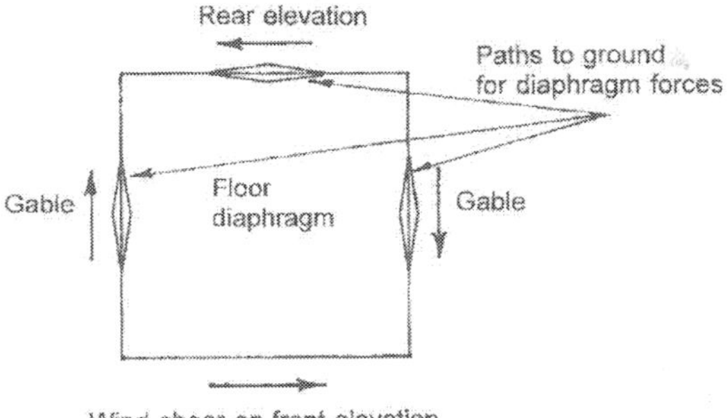

Figure 11. Statical system for carrying wind shear on the front elevation by diaphragm action

- The wall structure can, of course, also act as a diaphragm. With conventional stud walling, it is possible to take advantage of suitable board materials when they are properly fixed to the supporting members and some research has been carried out in order to determine the relevant limits. Without quoting their sources or the limitations, Grubb and Lawson (1997) specify "self-drilling self-tapping screws, or the equivalent, at a maximum spacing of 300 mm, although a spacing of 150 mm is more likely to be appropriate at the edges of the panels." It appears that this recommendation is based on tests carried out at the University of Surrey for British Steel (Griffiths and Wickens (1995). However, although interesting, these tests do not appear to justify the universal use of the above specification without the consideration of specific situations. Before using the diaphragm action of conventional walling, it is necessary to justify it by test or calculation in the light of the actual forces present. There does not appear to be a definitive design approach to this situation. Other authors report test results that demonstrate that there is a significant effect. However, none can be said to offer a definite design procedure.

- It is possible to use conventional diaphragm action with either a flat or trapezoidally profiled sheet metal lining although this may not always be economically practical. However, the author does know of one example where a flat sheet metal outer skin has proved to be successful in the construction of modular hotel room units.

- However, what is clearly practical is to use conventional diaphragm action in cassette wall construction. The special considerations in this case will be discussed later.

With all of the wall systems discussed above, it is expedient to space the holding down points as far apart as possible in order to reduce the shear forces in the wall and, at the same time, reduce the holding down forces into the foundations. This has implications for the architectural design so that early interaction between the Architect and the Engineer is required. The design of the holding down detail itself is another critical point in the structural design.

7.2 Stressed skin Design in Residential and Commercial Buildings

7.2.1 Introduction

Stressed skin action was originally discovered in the context of industrial buildings with pitched roof portal frames. As shown in Figure 1, under vertical load, the frames try to spread and this action is resisted by the roof sheeting and purlins which, together, act rather in the manner of a very deep (and therefore stiff) plate girder. Because of the proportions, considerations of shear predominate over bending. Notwithstanding this origin, most practical applications of stressed skin design have been concerned with flat roof construction, as shown in Fig. 2. The diaphragm action of the roof sheeting and its supporting members can be used to replace wind bracing in the plane of the roof with both a saving in material and a simplification of the detailing. In Fig. 2, the gables of the building are shown braced in order to provide a path to the foundations for the diaphragm forces. This gable may itself be sheeted and act as a stressed skin diaphragm, in which case the gable bracing may also be omitted.

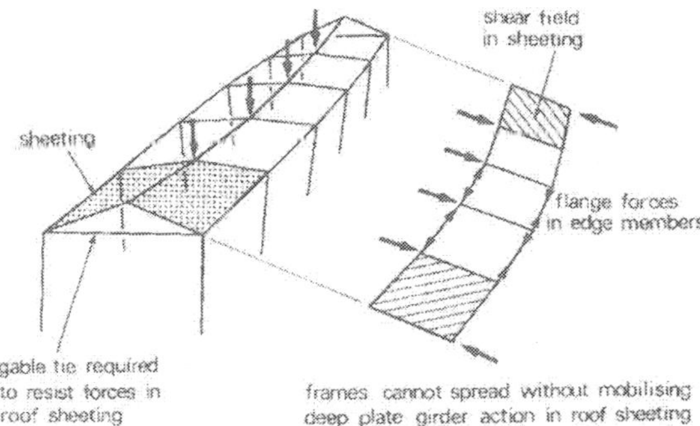

Figure 1. Stressed skin action in a pitched roof portal frame structure carrying vertical load

Early research into diaphragm action was concerned with buildings clad with a single skin of trapeziodally profiled metal sheeting. As the result of extensive testing and theoretical analysis over many years, a complete design procedure was produced (Davies and Bryan, 1981). The generic design document is the "European Recommendations" (ECCS, 1995) which were first published in 1977 and updated in 1995. This design procedure has now been incorporated into many National Codes of Practice (eg BSI, 1996) and enabling clauses are included in Eurocode 3: Part 1.3 (EC3, 1996). Recent research has extended consideration to a wide range of "modern" cladding systems such as two-skin built-up systems (Davies and Lawson, 1999).

The standard design element is the basic shear panel, a typical case of which is shown in Fig. 3. This case may be described as a "cantilever diaphragm" and includes the sheeting, the supporting members and the individual fasteners. There are a number of failure modes to be examined and it is frequently the fasteners which provide the ultimate limit state design criteria.

Design expressions are available for the strength and stiffness of both cantilever diaphragms and the assemblies of panels (often termed "diaphragm beams") shown in Figs. 1 and 2. These design expressions include many practical factors such the orientation of the sheeting, the arrangement of sheet lengths and openings for roof lights.

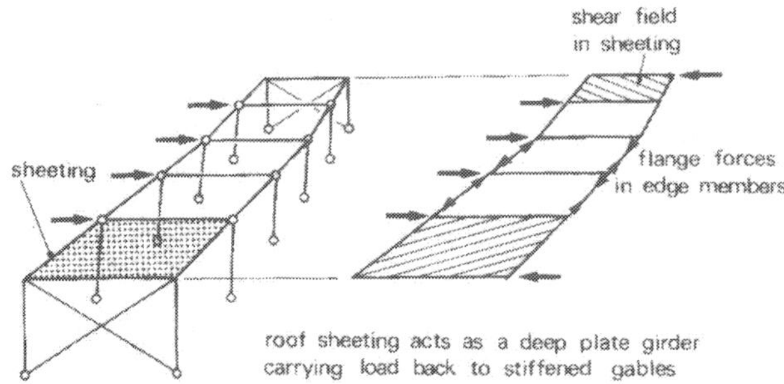

Figure 2. Stressed skin action in a flat-roofed structure with non-rigid frames

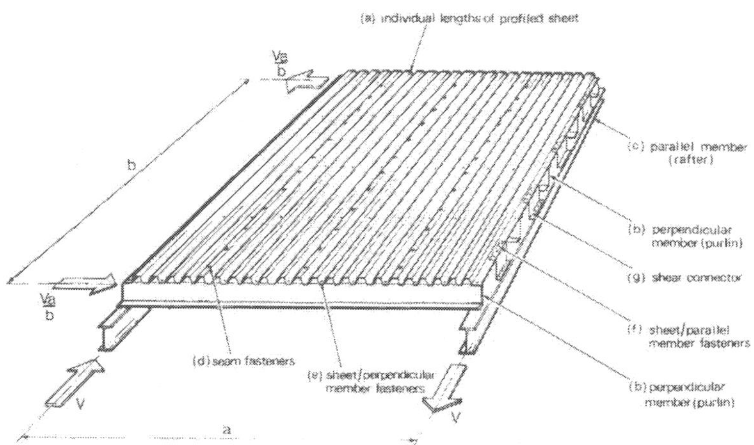

Figure 3. Basic arrangement of a cantilever diaphragm

Evidently, this whole scenario is directly applicable to the design of steel framed houses and similar constructions although it does not appear to have been widely used in this context.

7.2.2 Types of diaphragm

As discussed above, the two fundamental types of stressed skin diaphragm are:

- Diaphragm beams, typified by Figs. 1 and 2, in which the diaphragm spans between two gables or similar paths to ground for diaphragm forces.
- Cantilever diaphragms, typified by Fig. 3 in which a rectangular shear panel is loaded on one side and restrained on the other three.

Diaphragm beams arise more frequently in practice although cantilever diaphragms may be important in special situations. The great majority of the tests used to establish current stressed skin design theory have been carried out using cantilever diaphragms.

For both types of diaphragm, there are two sub types. Fig. 4 shows the case of the sheeting or decking spanning at right angles to the span of the diaphragm whereas Fig. 5 shows the alternative arrangement in which the sheeting or decking spans in the same direction as the span of the diaphragm. From the practical point of view, both arrangements give rise to similar levels of strength and stiffness although they have to be calculated differently.

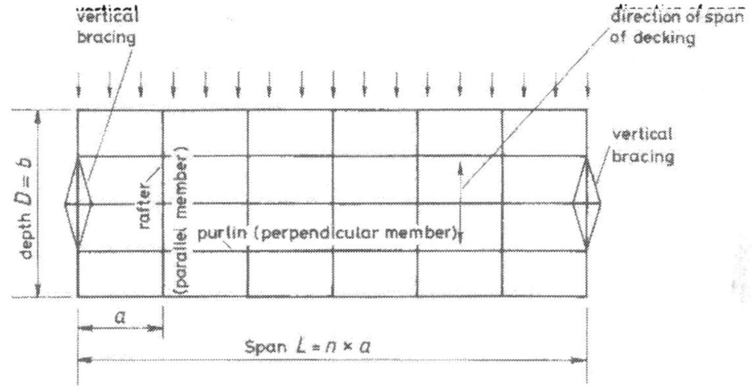

Figure 4. Diaphragm beam with decking spanning perpendicular to the span

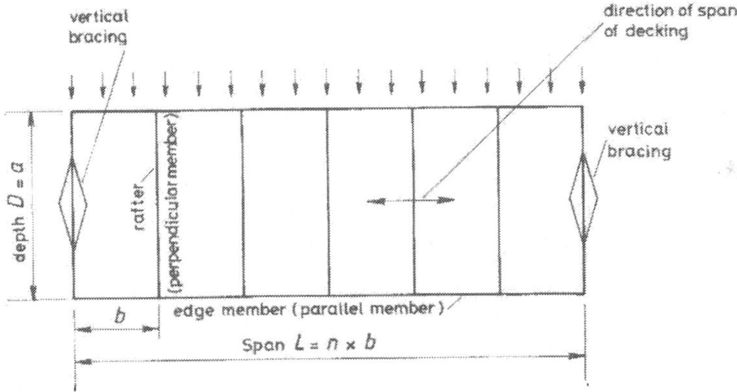

Figure 5. Diaphragm beam with decking spanning parallel to the span

Another sub-grouping of diaphragms is illustrated in Fig. 6. This concerns the way that they are fastened to the supporting structure. The normal procedures for fixing sheeting and decking result in diaphragms fastened on two sides only (indirect shear connection). Four side fastening (direct shear connection) is very advantageous but usually requires that special measures are taken. One possibility is to provide "shear connectors" in the form of purlin offcuts, as shown in Fig. 3.

With all of these arrangements, it is necessary to provide a path to ground for the forces in the diaphragms. With diaphragm beams, this usually takes the form of end gables to the building which are cross braced or otherwise stiffened, as illustrated in Figs. 1 and 2. With a cantilever diaphragm, three reaction forces must be provided, as shown in Fig. 3.

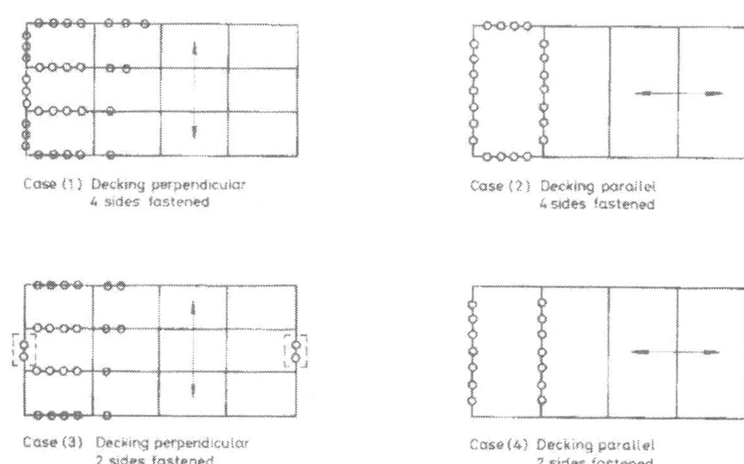

Figure 6. Arrangements of diaphragms with either 2 or 4 sides fastened to the supporting structure

7.2.3 Irregular roof shapes

Many buildings have an irregular plan form, as illustrated by Fig. 7. In such cases, it is necessary to divide the roof into a number of rectangular diaphragms and to ensure that each individual diaphragm has the required path to ground for the relevant forces. In providing this path to ground, it has to be appreciated that wind can blow from all sides! Thus, in Fig. 7, A and C are diaphragm beams whose extent is indicated by the diagonal lines. These are provided with a path to ground for wind blowing either up and down or across the page. D is a cantilever diaphragm for wind blowing from side to side, which requires a path to ground on three sides. B is an area that is not required to act as a diaphragm for wind blowing in any direction. This is as well because it contains rather a lot of roof light openings in a small area.

7.2.4 Calculation of diaphragm flexibility

Although structural engineers usually think in terms of stiffness, here it is logistically simpler to work in terms of flexibility, the reciprocal of stiffness, and this section briefly summarises the calculations of the flexibility of a complete diaphragm assembly such as shown in a number of previous figures. The design expressions embrace both beam and cantilever diaphragms and all sub-types.

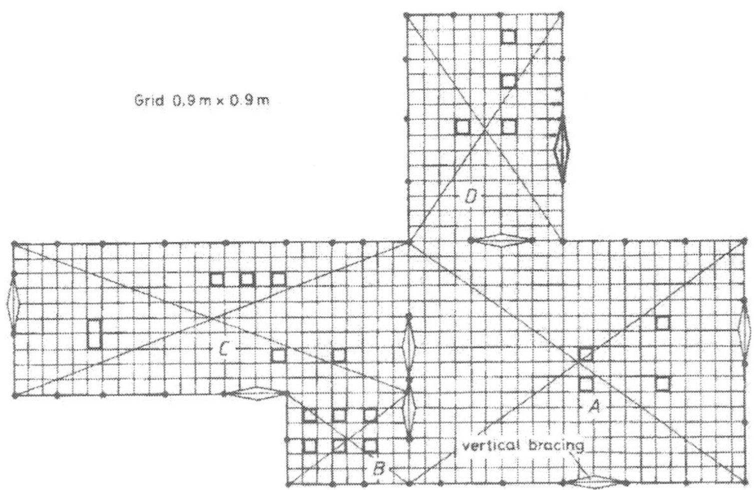

Figure 7. Irregular roof - division into diaphragm areas

Note: 1. No attempt is made here to derive design expressions. The derivations are covered in Davies and Bryan (1981). The design expressions themselves are all simple formulae, though requiring the use of tables. These are not given but are also available in all of the primary references.

2. No attempt is made to introduce secondary factors such as openings or sheet end laps within the diaphragm though it should be appreciated that techniques are available to deal with these.

The diaphragm flexibility `c' is defined as the shear deflection per unit shear load in a direction parallel to the corrugations in the sheeting (Fig. 8). For simple diaphragms (which embraces the majority of practical cases) this flexibility may be calculated by summing the following component flexibilities:-

$c_{1.1}$ = flexibility due to distortion of the corrugation

$c_{1.2}$ = flexibility due to shear strain in the sheeting

$c_{2.1}$ = flexibility due to movement in sheet to purlin fasteners

$c_{2.2}$ = flexibility due to movement in seam fasteners

$c_{2.3}$ = flexibility due to movement in shear connectors (or purlin to rafter connection in the case of direct shear transfer)

c_3 = flexibility due to axial strain in the purlins or secondary framing members

Simple expressions for each of these flexibilities are given in ECCS (1995), BSI (1996) and EC3 (1996) etc.

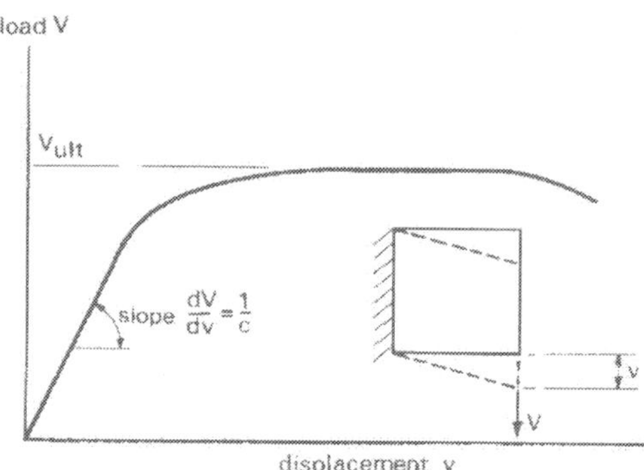

Figure 8. Typical load-deflection curve of a shear panel

Evidently, flexibility due to movement at the fasteners plays a major role. The design expressions require that the flexibility of the individual fasteners is known and this can be determined by a simple test on a lapped connection. Alternatively, conservative design expressions are available for the majority of the usual fastener types such as self-drilling, self-tapping screws, blind rivets and fired pins.

The other major component of total flexibility is $c_{1.1}$, the flexibility due to distortion of the corrugation. This complex issue has been the subject of a good deal of research and the design expression was first found empirically and later justified on a sounder theoretical basis as a result of the research described in Davies (1986). The design expression requires a constant 'K' which is a property of the cross-section. Values of this constant have been tabulated for trapezoidal and similar profiles fastened in either every or alternate corrugation troughs.

7.2.5 Calculation of the ultimate strength of a diaphragm

The diaphragm strength, V_{ult}, is defined as the ultimate capacity of the diaphragm under shear load as shown in Fig. 8. Failure of an individual diaphragm may occur in any one of a number of alternative modes, as follows-

1. Failure at seam fasteners between individual sheet lengths.
2. Failure at the sheet to purlin fasteners.
3. Failure at sheet to shear connector fasteners (direct shear transfer only).
4. Failure by buckling of sheeting in shear. This is unlikely in panels of sheeting or conventional gauge and typical fastener spacing.
5. Failure at the sheet ends due to collapse of the profile. This is relatively rare but can influence the design when long span decking profiles are used (Davies and Fisher, 1979).
6. Buckling of the apex purlin. Unlike 1 - 5, this will not occur in the end panel of the building but at the centre owing to bending forces as the sheeting panels form a deep plate girder spanning between the gables.

The strength of the diaphragm is given by the lowest failure load obtained when each of the above modes is considered. Again, fastener performance dominates the considerations with seam failure being the most common failure mode.

When designing diaphragms, it is always good practice to design so that the failure is ductile. This means that the diaphragm should always be designed to fail in one of the following modes:-

Diaphragm fastened on four sides (direct shear transfer):-	failure at seam fasteners or failure at shear connector fasteners
Diaphragm fastened on two sides (indirect shear transfer):-	failure at seam fasteners, or failure at end sheet to purlin fasteners

It is recommended that other failure modes should show a 25% reserve of safety relative to the lowest load associated with the above modes. Simple expressions for the calculation of failure loads are available in the literature.

7.2.6 Interaction of diaphragm action and rigid-jointed plane frames

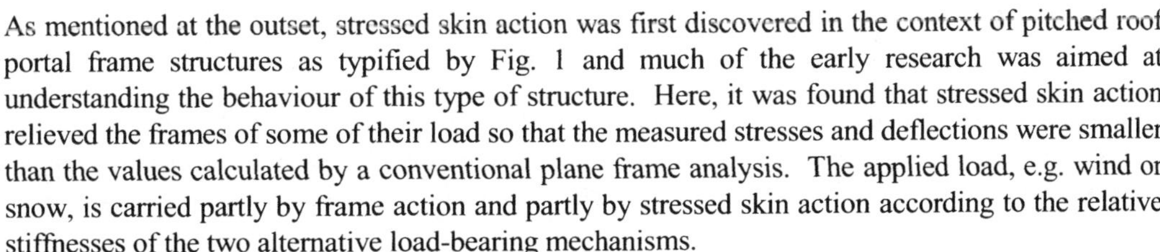

As mentioned at the outset, stressed skin action was first discovered in the context of pitched roof portal frame structures as typified by Fig. 1 and much of the early research was aimed at understanding the behaviour of this type of structure. Here, it was found that stressed skin action relieved the frames of some of their load so that the measured stresses and deflections were smaller than the values calculated by a conventional plane frame analysis. The applied load, e.g. wind or snow, is carried partly by frame action and partly by stressed skin action according to the relative stiffnesses of the two alternative load-bearing mechanisms.

Nowadays, the roof slopes of pitched roof portal frames are rather low, typically of the order of 5°, and in such cases it is found that the contribution of stressed skin action to the overall behaviour under vertical (snow) load is rather small. As this load case tends to govern the design of the majority of pitched roof frames, interest in this aspect of stressed skin design has waned. Conversely, however, stressed skin action is much more effective in helping to control sway under wind blowing from the side. Indeed, in tall sway frames, a high proportion of the side load is likely to be carried by stressed skin action, whether or not the designer wants this. These considerations have a strong influence on the practical usage of stressed skin design.

7.2.7 Suitable structures for stressed skin design

Stressed skin action reduces the deflections of the joints of the structure in the planes of the cladding. It does not reduce the deflections within the lengths of the members normal to the planes of the cladding, (i.e. the `no sway' components of bending moment and deflection are unchanged), nor does it have a significant effect at joints where members meet at an angle close to 180°. It follows that typical situations where stressed skin construction may advantageously be employed are:-

(a) In flat roof construction, all of the side load may be carried by diaphragm action and transmitted to ground by suitable vertical bracing. This has been by far the most important use of stressed skin design to date in both `system' built and prestigious `one-off' construction;

Note: In such situations, diaphragm action can also provide lateral restraint to beams.

(b) In relatively tall frames, particularly those containing cranes, where the lateral deflection at eaves level is critical or where relative lateral movement between the gable frame and the intermediate frames causes considerable forces in the cladding whether or not this is considered in the design of the frames;

(c) Stressed skin action provides a convenient means of carrying gable wind forces to ground via the roof and side cladding without the need for diagonal bracing.

(d) In tied portal frames, where the design of the stanchions is governed by side load considerations.

(e) In structures subject to a lateral point load. Stressed skin action is particularly effective in distributing such a point load among adjacent frames.

It should be noted that, as a general guide, unless the diaphragm is very lightly loaded, deflections are likely to be excessive or the benefit of stressed skin action small if the ratio of span to depth of a diaphragm exceeds about four.

7.2.8 The danger of ignoring stressed skin action in conventional construction

Regardless of whether the design of a clad building is elastic or plastic and whether the design takes into account the stiffening effect of cladding or not, it is likely that the relative strengths and stiffnesses of the frames and sheeting panels are such that, under increasing load, first yield will take place in the sheeting panels. This is particularly the case in relatively tall, steel-framed structures where a major design factor is lateral loading due to wind and where the relative displacement between the gable frame and adjacent frame can seriously embarrass the cladding. In such structures, yield or even failure of the sheeting panels can occur at the working loads. As has been pointed out earlier, stressed skin action is present whether the designer acknowledges it or not. The methods of stressed skin design allow an analysis to be made of such cases so that the forces in the sheeting may be investigated. The author believes that this is the correct way to proceed whether or not it is intended to take account of stressed skin action in the design of the frame.

A number of tall rectangular buildings have been analysed in order to investigate the condition of typical roof sheeting panels at the working loads. One of these theoretical investigations of an actual building some 15 metres high has been reported in some detail (Davies and Bryan, 1981). It is shown that tearing of the sheeting at the fasteners may be predicted at wind loads well below the working values. As the structure was designed on the basis of simple sway frames, this does not suggest total failure but it could well indicate difficulty in keeping the building watertight. This is a typical pattern of behaviour in tall, rectangular framed buildings and is particularly critical when the cladding is fastened to the supporting members on two sides only, which is invariably the case when no consideration of stressed skin effects is included in the design. In such structures, calculated sway moments and deflections are likely to be completely fictitious and the forces in the cladding panels large enough to demand serious consideration.

7.2.9 The application of stressed skin design to typical house construction

The considerations governing the stability of a typical house, or similar low rise construction, have been summarised in section 7 of the first part of this chapter and, in particular, Fig. 7. In general, one or the other of bracing or stressed skin action are required in the floors, roof and walls.

7.2.9.1 Stressed skin action in floors

In buildings with light steel framing, the floor construction is logically lightweight and dry. Therefore, it often takes the form of a timber-based product supported by steel joists as shown in Fig. 4 of part 1. It is sometimes assumed that conventional timber-based floor construction can provide the necessary diaphragm action without formal calculations. This is a highly dangerous assumption! Diaphragm action in the floor is fundamental to the stability of the structure and the in-plane forces can be very significant. The situation is not helped by the presence of a stair well. The diaphragm action of the floor must always be properly calculated when it will be found that the main problem is in providing the required shear connection between the timber floor and the primary structure at the same time bearing in mind the absence of any 'seam strength' between the individual sheets of floor material. The basic principles of stressed skin design, as described above, can be applied to this form of floor construction. More details of the procedures for this calculation may be found in the literature on timber construction!

With light steel framing, a more reliable solution is to provide a light gauge steel floor deck as shown in Fig. 5 of part 1. This allows the formal calculation procedures of stressed skin design, to be used. When these procedures are properly used, it is found that the required shear strength can be readily achieved, helped by the provision of seam fasteners. The demands made on the relatively reliable metal to metal fasteners in this form of construction highlight the potential difficulties with timber floors.

Evidently, steel-concrete composite decks are an alternative form of construction which can readily meet all of the design criteria of structural performance and building physics. Davies and Fisher (1979) discuss the diaphragm action of composite deck slabs.

7.2.9.2 Stressed skin action in roofs

This is conventional stressed skin action and no new considerations are introduced.

7.2.9.3 Stressed skin action in walls

There are several methods available for providing in-plane stability in vertical walls and these have been summarised in section 7 of part 1. From the point of view of stressed skin design, the most interesting is the use of cassette wall construction which will be discussed in detail in the next section.

7.2.10 Current practical applications of stressed skin design (Davies and Bryan, 1981 etc)

In the early days of the development of stressed skin theory, it was widely used to remove the need for wind bracing in the plane of the flat roof in system built schools, offices etc. The intention was not only to save the cost of the bracing but, at the same time, to simplify the system (and its detailing) by reducing the number of components. Indeed, design tables were provided so that the architect could verify the stability without the need for any structural engineering input. Subsequently, it has been extensively used in a similar fashion to eliminate the wind bracing in flat-roofed industrial buildings.

As discussed in other parts of this series of papers, cassette wall construction, as utilised in the "CIBBAP" system makes extensive use of stressed skin action. Modular construction, with cassette walls, or indeed with other types of metal skin wall, is another application which is often fundamentally dependent on stressed skin action.

In the author's experience, it arises from time to time in "special" structures, generally to remove the need for wind bracing, the Commonwealth Stadium in Manchester being a recent example. Finally, there are a few examples of folded plate and "frameless" light gauge steel construction to be found in practice.

7.3 Cassette Wall (Structural Liner Tray) Construction

7.3.1 Introduction

Light gauge steel cassettes (also known as structural liner trays) offer an alternative form of wall assembly for use in low-rise steel-framed construction. The basic arrangement of a cassette wall is shown in Fig. 3 of part 1. Lipped C-shaped cassettes span vertically between top and bottom tracks to form panels which may be storey height or higher. When used in this way, the narrow flanges may point towards either the inside or the outside of the building. The preferred arrangement is for the narrow flanges to point inwards so that the wide outer flange immediately provides a weatherproof membrane and no second steel skin is necessary. The wall construction is then completed internally by insulation and a dry lining, an option often referred to as "cold-frame" construction. The alternative, with the insulation and a waterproof outer skin external to the cassette wall is "warm-frame" construction.

The idea of cassette wall construction appears to have originated from Baehre in Stockholm in the late 1960's (Thomasson, 1978). He envisaged the steel cassette as the primary structural element in modular construction as illustrated in Fig. 1. When used in this way, as shown in Fig. 2, cassette walls are subject to the three primary load systems of axial compression (from the storeys above), bending about the minor axis (from wind pressure and suction) and shear (from wind-induced diaphragm action). Accordingly, Baehre's research team investigated in some depth the performance of cassette sections when subject to each of these three actions individually (Thomasson, 1978; Konig, 1978; Nyberg, 1976).

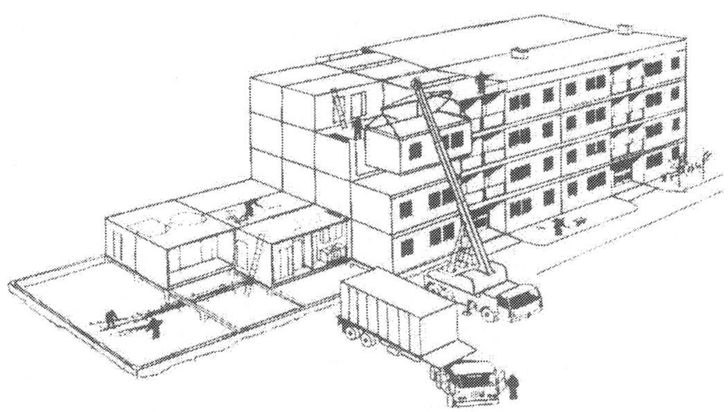

Figure 1. Modular construction with light gauge steel cassettes

Cassette construction may be viewed as being similar to conventional light gauge steel stud construction integrally combined with a metal lining sheet to provide a metal frame together with a flat steel wall. Its main advantages, in comparison with the alternative stud wall system, may be summarised as follows:

- Simple details and rapid construction

- The wall structure is immediately water-tight
- The stability problems of thin slender studs are avoided
- A rational provision for wind shear can be made without additional bracing members by utilising stressed skin action

The last bull point is particularly important and is discussed in some detail later.

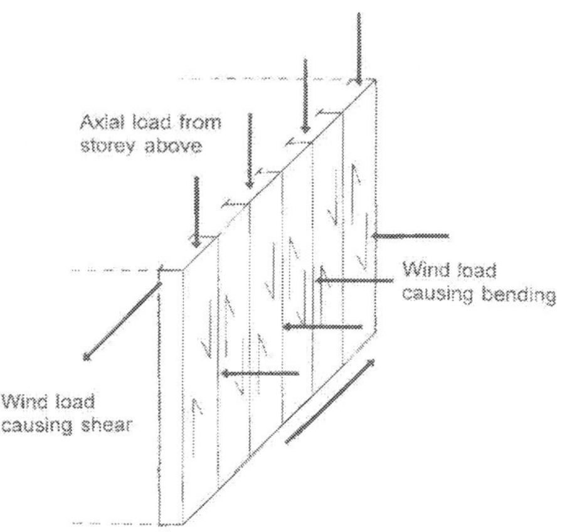

Figure 2. Load system in a cassette wall

As far as the author is aware, Baehre's ideas were not put into action until much later when the "Scanmodule" system (Davies et al., 1995) used the construction method illustrated in Fig. 1. However, the cassette wall system has recently been pioneered in France by the company 'Produits Acier Batiment' (PAB) under the name 'CIBBAP' and more than 20 projects have been completed using this concept.

In recent years, cassette sections have also been widely used in an alternative wall construction in which they span horizontally between structural frames. Here the cassettes interact with and are stabilised by a relatively light trapezoidally profiled metal outer skin which spans at right angles to the span of the cassette. For both thermal and acoustical reasons, the troughs are usually filled with insulating material. When used in this way to form a two-layer, built-up cladding system, cassettes are often termed "structural liner trays". Typical construction of this type is shown in Fig. 3. A similar arrangement to Fig. 3 is also used in roof construction.

Baehre, by this time Professor of Steel Construction at the University of Karlsruhe in Germany, then turned his attention to structural liner trays used in this way and carried out a significant number of tests (Baehre and Buca, 1986; Baehre, 1987; Baehre et al., 1990). These formed the basis of the design clauses in Eurocode 3 : Part 1.3 (EC3, 1996) which will be referred to as "EC3". As these design procedures are based on test results, EC3 places geometric restrictions on cassettes that appear to reflect the limits of Baehre's tests rather than any fundamental restrictions of the structural system.

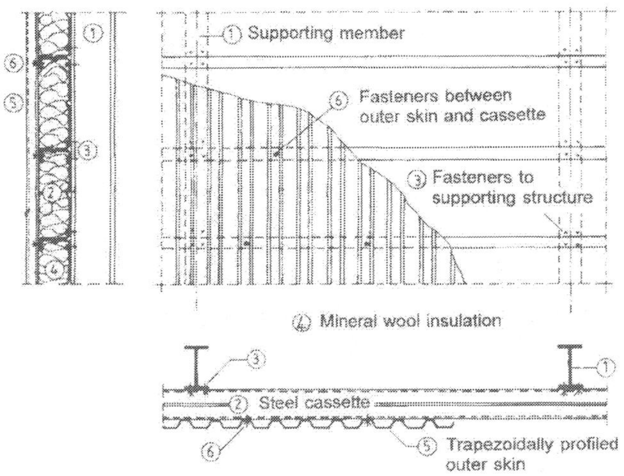

Figure 3. Two-layer built-up wall cladding system

Fig. 4 shows the geometry of a typical cassette together with the symbols used in EC3. These symbols will also be used in this paper. The elements of the section are described in this paper as the wide flange, the webs, the narrow flanges and the lips. The range of validity of the design procedures in EC3 is stated to be as follows:

0.75 mm	$\leq$	t_{nom}	$\leq$	1.5 mm
30 mm	$\leq$	b_f	$\leq$	60 mm
60 mm	$\leq$	h	$\leq$	200 mm
300 mm	$\leq$	b_u	$\leq$	600 mm
		I_a/b_u	$\leq$	10 mm^4/mm
		h_u	$\leq$	h/8
		s_1	$\leq$	1000 mm

Where I_a is the second moment of area of the wide flange about its own centroid as shown in the right hand insert on Fig. 4.

It may be noted here that, primarily because of "flange curling" which is discussed later, there is more scatter in the comparison of test results with theory than is customary with cold-formed sections in bending. The Eurocode therefore requires a material factor of $\gamma_{M2} = 1.25$ for bending strength in contrast to the more usual values of $\gamma_M = 1.0$ or 1.1. It is implicit, therefore, that significantly better bending strengths are likely to be obtained by test than by calculation and, for a mass-produced product, this is recommended.

As the clauses in EC3 explicitly require the stabilising effect of the second metal skin, it is necessary to reconsider their applicability to cassette wall construction where this second skin is unlikely to be present (or may be replaced by a much weaker material such as plasterboard).

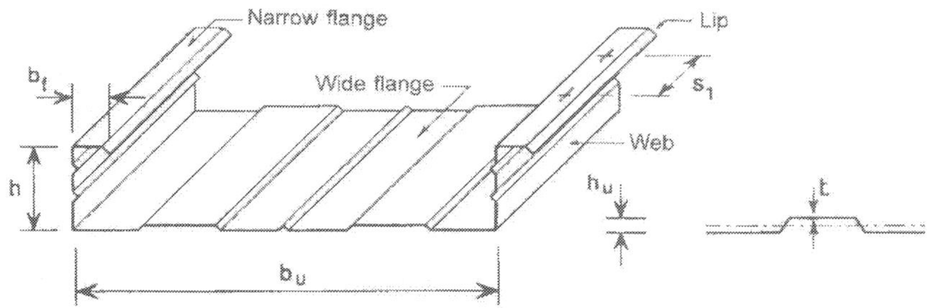

Figure 4. Typical cassette section

7.3.2 Design procedures for cassettes

In the following sections, each of the basic load systems shown in Fig. 2 is considered in turn in the light of Baehre's research and the resulting clauses in EC3 (1996) bearing in mind the absence of the second skin which is assumed in the EC3 procedures.

7.3.2.1 Axial compressive load

This load case is not explicitly considered in EC3. However, it is implicit that the cross-section model shown in Fig. 5 should be used in which the wide flange and web are treated by conventional effective width procedures and the narrow flange by the more complicated procedures for a flange stiffened by a lip. This requires an iterative procedure in order to determine the reduced effectiveness of the combination of the lip and the outer portion of the flange. This part of the design procedure for the axially loaded cassette is similar to that for bending with the narrow flanges in compression, which is given in some detail in EC3. The remainder of the design procedure then follows that for any thin-walled column bearing in mind that it is completely stable with respect to buckling in the plane of the wall. It is also necessary to consider distortional buckling (tripping) of the narrow flange. The considerations with regard to distortional bucking are similar to those considered in the next section.

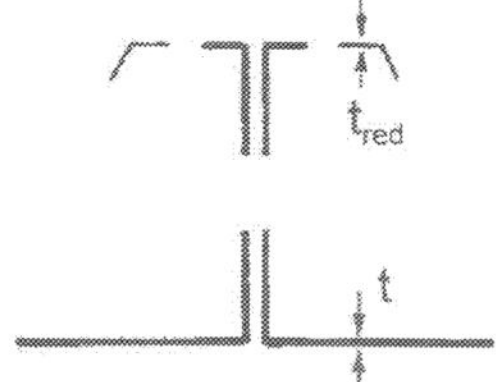

Figure 5. Design model for a cassette subject to axial compression

7.3.2.2. Behaviour in bending with the narrow flange in compression

The behaviour of cassette sections in bending is characterised by all of the usual considerations of thin-walled, cold formed section construction with the addition that the wide flange tends to curl towards the neutral axis, as illustrated by Fig. 6, whether this flange is tension or compression. The available analytical procedures for dealing with "flange curling" are far from rigorous and the procedures in EC3 are largely empirical. Consequently there is a good deal of scatter of test results and a degree of in-built conservatism.

The design for bending with the narrow flange in compression is a particularly complicated design problem because there are three effects to consider:

- local buckling of the web and narrow flange.
- distortional buckling (tripping) of the narrow flange/lip combination
- "flange curling" of the wide flange (which is in tension)

The first and last of these are considered in detail in clauses 10.2 and 10.3 of EC3. Flange curling is treated by an effective width approach. In cassette wall construction, the absence of the second skin of cladding will not have any influence on these aspects of behaviour.

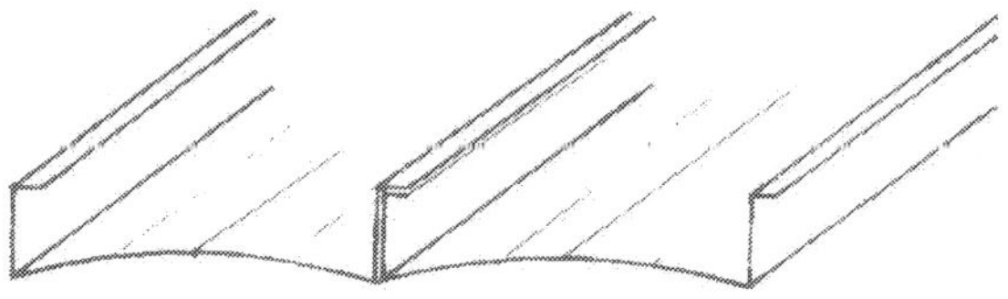

Figure 6. Flange curling

However, distortional buckling is more problematical because the minimum fastener spacings in EC3 clause 10.2.1 are clearly intended to prevent this phenomenon. We may note, however, that the requirements are very simplistic and presumably reflect the observation that distortional buckling was not critical in any of Baehre's test panels when this fastener specification was present. It does not automatically follow that these fasteners to an auxiliary bracing system are always necessary. What is required is a separate check for the distortional buckling stress of the unbraced cassette. It may also be born in mind that the web connections between adjacent cassettes, which are usually necessary in cassette wall construction, will also inhibit distortional buckling.

Procedures for determining the distortional buckling stress of this type of section have been discussed elsewhere. Accurate solutions are available with the aid of second-order Generalised Beam Theory (Davies et al., 1994). The simplified design models proposed by Serrette and Pekoz (1995) are also available for this problem. It appears that, in many cases of cassette wall construction, no reduction of the narrow flange compressive stress is required in order to cater for the possibility of distortional buckling.

7.3.2.3 Bending with the wide flange in compression

Here, the more complex narrow flange and lip assembly is in tension and does not buckle. The bending behaviour is, therefore, dominated by local buckling of the wide flange. However, flange curling as described above also occurs when the wide flange is in compression and interacts with local buckling. EC3 does not propose any rigorous treatment of this interaction and, indeed, this would appear to be exceedingly difficult. Instead, it suggests that the beneficial effect of intermediate stiffeners should be neglected and that the conventional effective width procedure should be used though with the material factor γ_m increased to 1.25 in order to deal with the additional uncertainty caused by flange curvature. The absence of the second skin clearly has no influence when the narrow flange is in tension.

7.3.2.4 Behaviour in shear

A cassette sub-assembly is also a ready-made shear panel or `diaphragm' for stressed skin construction as discussed in part 2 above. Stressed skin design is explicitly allowed in EC3 and appropriate enabling clauses are included in section 9. EC3 also includes somewhat rudimentary provisions for cassettes acting as shear panels. These make it clear that the behaviour of a cassette wall panel in shear is not significantly different from that of a conventional shear panel comprising trapezoidal steel sheeting framed by appropriate edge members so that the procedures described in the definitive publications (Davies and Bryan, 1991; ECCS, 1995) may be used.

There are three main differences between cassette and liner tray systems and the trapezoidally profiled roof sheeting and decking for which the calculation procedures were originally devised:

- There is negligible flexibility due to shear distortion of the profile. This removes a design equation which tends to dominate the deflection calculation for trapezoidal profiles. Here it is possible to make a simplified estimate of deflections based on the assumption that the flexibility arises mainly in the fastenings.
- The strength calculation tends to be dominated by the tendency of the wide flange to buckle locally in shear before any of the more usual diaphragm failure modes (fastener failure, profile end failure or global shear buckling) are mobilised.
- There is often no separate edge member parallel to the cassettes. This means that there are no longitudinal edge fasteners to check and the web and narrow flange of the outermost cassette act as their own edge member which should be checked for the induced compressive force.

The first two of these considerations lead to the two equations given in EC3 for the ultimate and serviceability limit states respectively. These equations may be derived from the more familiar equations as follows:

Baehre (1987) shows that the simplified Easley equation is valid for the determination of the shear flow $T_{v,Sd}$ to cause local buckling:-

$$T_{v,Rd} = \frac{36}{b_u^2}\sqrt[4]{D_x D_y^3}$$

where D_x = bending stiffness across the wide flange $\approx \dfrac{EI_x}{b_u}$

$$D_y = \text{bending stiffness along the wide flange} = \frac{Et^3}{12\left(1-\nu^2\right)} = \frac{Et^3}{10.96}$$

I_a = second moment of area of the wide flange about its own centroid.

Thus:
$$T_{v,Rd} = \frac{36E}{\sqrt[4]{10.96^3}} \sqrt[4]{\frac{I_a t^9}{b_u^9}} = 6E \sqrt[4]{I_a \left(\frac{t}{b_u}\right)^9}$$

This is the equation in clause 10.3.5 of the Eurocode. It should be noted here that this equation is printed incorrectly in both Baehre's paper and in the original printing of the Eurocode. It is corrected in the Corrigenda to ENV 1993-1-3 dated 1997-02-25.

For the calculation of deflection, Baehre (1987) considered the following four components of shear flexibility (Davies and Bryan, 1991; ECCS, 1995):

- $c_{1.2}$ shear strain
- $c_{2.1}$ flexibility of the sheet end fasteners
- $c_{2.2}$ flexibility of the seam fasteners
- $c_{2.3}$ flexibility of the "shear connector" (longitudinal edge) fasteners

He showed that all are of similar magnitude and stated (without further explanation) that their resultant stiffness could be approximated by:

$$S_v = \frac{2000Lb_u}{E_s\left(b-b_u\right)} \quad \text{(N/mm)}$$

which is the equation to be found in clause 10.3.5 of the Eurocode.

This rather crude approach to stiffness appears to be justified because, in the absence of the distortion term, the deflections tend to be small and because cassettes tend to be of fairly similar proportions and to have similar fastening systems, the individual fasteners of which have similar flexibilities. However, this simplification is not essential and, if the deflections are at all critical, the more fundamental approach to the calculation of deflections given in Davies and Bryan (1991) and ECCS (1995) is to be preferred.

More importantly, the wording of the clauses in EC3, and the above equation for the shear flow to cause local buckling, may lead designers to overlook that fastener strength may also be critical. In addition to considering local buckling of the wide flange, it is essential also to consider the possibility of failure in each of the fastener failure modes considered in conventional stressed skin theory, namely:

- failure in the seam fasteners between adjacent cassettes
- failure in the fasteners connecting the ends of the cassettes to the foundation or the primary structure
- failure in the shear connector (longitudinal edge fasteners)

It appears clear to the author that the excessive simplicity in the approach in EC3 to the design of cassettes subject to diaphragm action is likely to lead to a lack of fundamental understanding and over-confidence. The inevitable result will be serious design errors.

For a typical vertically-spanning cassette wall, at first sight, it might be thought that providing a continuous connection to the foundations at every seam line might reduce the forces in the vertical edge members and thus avoid high local uplift forces at the ends of the diaphragms. However, the following simple calculation carried out for an individual cassette with reference to Fig. 7 demonstrates that this is not the case:

$$P_1' = \frac{P'H}{b_u} = \frac{PH}{nb_u} = \frac{PH}{B_u}$$

It follows that continuous connection to the foundations merely has the potential to reduce the forces in the seam fasteners between adjacent cassettes but has no influence at all on the critical forces at the ends of the diaphragm. Indeed, resisting the uplift forces at the ends of the diaphragm walls is one of the more troublesome aspects of cassette wall design and there is scope for some ingenuity in achieving good practical details.

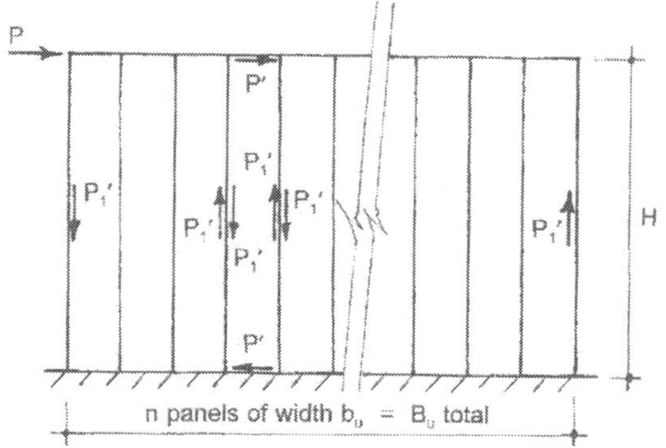

Figure 7. Diaphragm forces in cassette wall construction

It is expedient to space the holding down points as far apart as possible in order to reduce the shear forces in the wall and, at the same time, reduce the holding down forces into the foundations. This has implications for the architectural design so that early interaction between the Architect and the Engineer is required. The design of the holding down detail itself is another critical point in the structural design.

Fig. 8 shows an elevation of a cassette wall as used in a typical house facade. The lines x-x show the division into prefabricated sub-panels for factory construction. The wind-shear diaphragms are indicated by cross-hatching.

More recently, it has become clear that it is often better to treat a cassette wall with openings as a whole, as shown in Fig. 9, and to provide holding down points at the corners only. This reduces both the in-plane shear forces and the uplift forces on the foundations. Shear transfer across the window and door openings is via the roof and floor edge beams and the (concrete) ground beam. In the present state-of-the-art, this requires a finite element analysis but research is in progress to determine a more appropriate simplified design procedure.

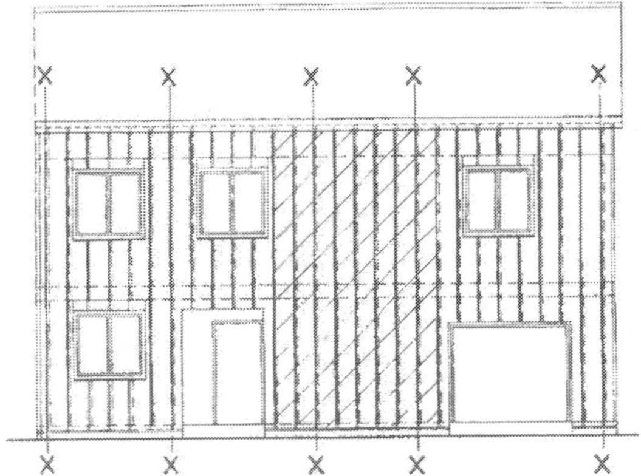

Figure 8. Typical cassette wall in house construction

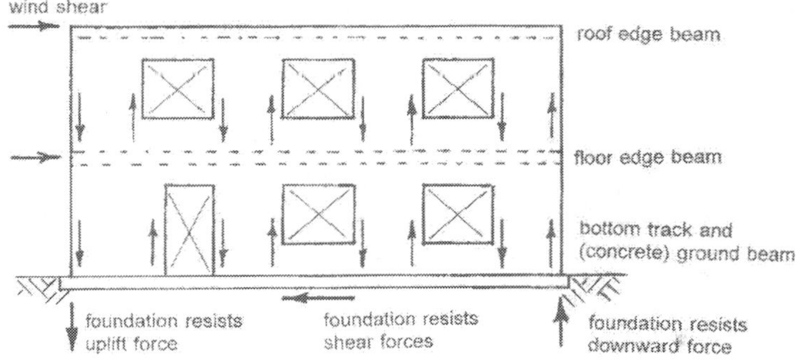

Figure 9. Improved treatment of diaphragm action in a cassette wall

7.3.2.5 Combined effects

Although considered separately in sections 3.2.1 to 3.2.4 of this paper, axial load, bending and shear effects may evidently interact. Noting that there is no major axis bending or buckling, minor axis bending and axial load can be readily combined according to clause 6.5.1 of EC3.
The shear stresses in stressed skin action are very low and it is usual to neglect interaction between in plane shear and primary axial load and bending. However, the axial compressive forces arising from stressed skin action in the edge of a diaphragm (see Fig. 7) must be combined with the axial force arising from load from the floors and roofs above. This will usually be of major design significance.

Cassette wall panels can, therefore, be readily designed on the basis of EC3 together with the established procedures for stressed skin design and it is found that, for most low-rise construction, a standard panel and fastener specification is sufficient to carry the wind shear without any special provision other than for holding down forces at the leeward end of the diaphragm.

7.3.3 Connections

Connections between individual cassettes and between cassettes and other structural elements may be made with any of the usual mechanical connections such as bolts, screws or blind rivets. Cassettes are substantial structural members and site connections are usually made by bolting. However, when forming prefabricated panels in the factory, a particular advantageous possibility is the use of press-joined connections (Davies et al., 1996) to form the seams between individual cassettes. Press-joining is quick, cheap and does not destroy the galvanising. The structural details tend to be simpler than those with stud construction so that erection is rapid.

The use of welding is not recommended as it destroys the galvanising leaving the wall more susceptible to corrosion damage.

7.3.4 Infilled cassettes

Cassettes are usually infilled with loose insulation. However, many insulation products, such as polyurethane, polystyrene and mineral wool, have inherent rigidity. When these materials are bonded to the thin steel, they increase the resistance to local buckling and site damage. Because of the slenderness of the wide flange, the increased resistance to local buckling under both axial compression and shear may be considerable.

At the present time the design procedures required to take advantage of this effect require further research. However, preliminary investigations which reveal its magnitude may be found in Davies and Hakmi (1991) (axial stress) and Davies and Dewhurst (1997) (shear).

7.4 Some Additional considerations in Light Gauge Steel Framed Construction

7.4.1 Introduction

This part of the series of papers discusses some of the more general consideration that arise when using cold-formed steel members as primary components in building construction. It draws heavily on previous papers written by the author, in particular Davies (1999).

7.4.2 Wall studs in steel framing systems

The primary building block of most steel framing systems is the wall stud. The author has long proclaimed that thin-walled, cold-formed steel is often at its best when interacting with other materials. This is particularly true in the case of wall studs where interaction with the lining (sheathing) material can substantially increase the load carrying capacity of the stud.

Wall studs generally have two and possibly three separate functions. Their primary function is to carry vertical load from the floors and roof above. In external walls, they also have to resist the lateral pressure from the wind and transmit this to the floor and roof diaphragms and to the foundations. In addition, certain studs may also form part of the system resisting in-plane forces from wind shear. Wall studs are typically C-shaped galvanised cold-formed sections with dimensions of the order of $100 \times 50 \times 1.5$ placed with their flanges in contact with the wall surface. The wall material may be, for example, a gypsum or wood fibre based board or plywood and, if this material has adequate strength and stiffness and if there is adequate attachment to the studs, then the axial load bearing capacity may be substantially increased by the resulting structural interaction. This is mainly as a consequence of the resistance provided against lateral buckling modes.

Evidently, wall studs can be designed as free-standing members without taking advantage of the influence of the other elements of the wall construction and modern codes of practice allow this on the basis of calculations alone. Inclusion of the stiffening influence of the walls has to be semi-empirical based on the interpretation of test results.

This effect has been included in the American design code (AISI, 1996) for many years and Yu (2000) outlines the historical development. The first tests date back to the 1940's and these showed that, in order for the necessary support to the stud to exist, the assembly must satisfy three requirements:

1. The spacing between attachments must be close enough to prevent the stud from buckling in the direction of the wall between attachments.

2. The wall material must be rigid enough to minimise deflection of the studs in the direction of the wall.

3. The strength of the connection between the wall material and the stud must be sufficient to develop a lateral force capable of resisting bucking of the stud without failure of the attachment.

The first AISI provisions were based on these findings. Subsequent research in the 1970's indicated that the bracing to the studs provided by the wall panels was of the shear diaphragm type rather than the linear spring type assumed in the earlier study. The AISI clauses were modified in 1980 to reflect these research findings. After some minor modifications in 1989, the current state of the art was defined in 1996 when extensive revisions were made to permit the use of wall studs with either solid or perforated webs (to permit the passage of services) (Miller and Pekoz, 1994). Yu (2000) gives a detailed example of the calculation procedure according to the AISI code.

Even more interesting is the use of studs (and tracks) with arrays of longitudinal slits in their webs as shown in Fig. 1. The purpose of such slits is, of course, to substantially reduce thermal bridging. Some manufacturers claim that their slotted studs are more thermally efficient than a timber equivalent. The one illustrated is said to have only slightly lower thermal performance than a wooden stud. Kesti and Makelainen (1998) have considered the design of this type of stud and have shown that, by reducing the transverse bending stiffness of the web, the slits make them particularly susceptible to distortional buckling. Taking into account the restraining influence of gypsum sheathing increases the design capacity but the sheathing cannot fully overcome the deleterious influence of the slits. Because of the more onerous requirements for thermal insulation, slotted studs have originated in Scandinavia and most of the development work has taken place in the Nordic countries. Hoglund (1998) summarises some of this research and gives further details of the appropriate design procedures.

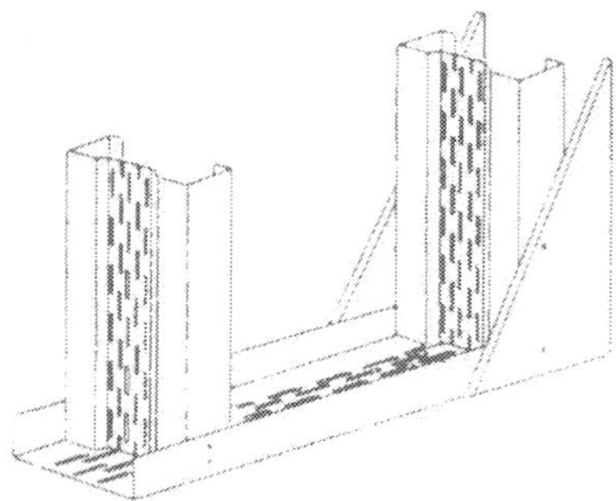

Figure 1. Wall construction with slotted thermal studs (Kesti and Makelainen, 1998)

Perhaps the ultimate wall element is the cold-formed steel cassette, filled with rigid insulation material and having slotted webs. To the best of the author's knowledge, this does not exist in practice and it is certainly worthy of development.

7.4.3 Connections for light gauge steel framing systems

In cold formed steel frame construction, a significant proportion of the structural cost is in the connections, with bolts being particularly expensive. Welds are also to be avoided because they destroy the protective galvanised coating, which is one of the fundamental advantages of the cold formed steel section. Attention therefore must be paid to developments in connection technology, particularly those methods which do not require the use of expensive components.

Evidently, any of the conventional methods of joining cold formed steel members together can be used in steel framed house construction and, despite their disadvantages, bolting and welding feature prominently in current design solutions. However, a relatively new technique, imported from the motor industry, known a press-joining can advantageously be used, particularly when panels are prefabricated in the factory.

Press joining is a single-step process which requires a tool consisting of a punch and expanding die as shown in Fig. 2. The tool parts have a rectangular profile and the punch tapers along the line of cut. The die has a fixed anvil with spring plates on either side. Although the process is a single action, it consists of two phases. In the first phase, the punch moves towards the die and forms a double cut in the two steel sheets. In the second phase, the pressed area is flattened against an anvil in order to spread the pressed out strip laterally and form a permanent connection.

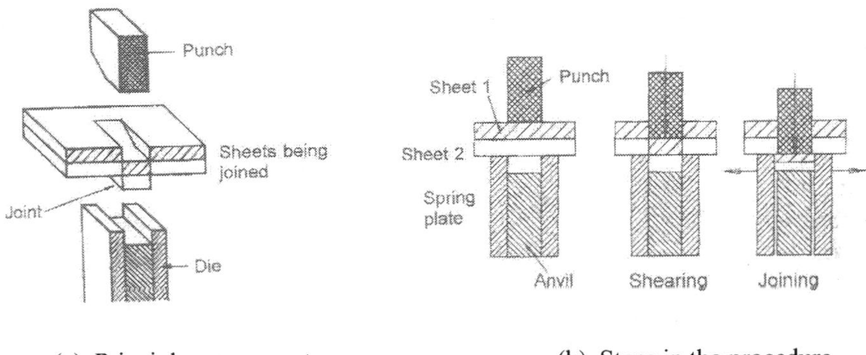

(a) Principle components (b) Steps in the procedure

Figure 2. Press jointing

The advantages of press-joining in the context of construction are (Davies et al, 1996):

- The join is formed using the material of the sheets to be connected and no additional items are required.
- It does not destroy protective coatings such as galvanising
- It is very rapid, taking less than one second to form a joint
- It is very energy efficient, requiring about 10% of the energy for spot welding
- Multipoint jointing tools can produce several joints simultaneously
- The joint can be made air and watertight

Davies at al. (1996) have discussed the strength and stiffness of press joins and have shown that these are entirely suitable for structural purposes. Helenius (2000) has given a detailed discussion of their mechanical strength and Davies et al (1995) have described their successful application in modular construction. Joins of this type have been extensively used in a number of buildings in the 'CIBBAP' cassette wall project in France. Evidently, this is a technique that should have a prominent place in the panel or modular construction of steel framed houses.

The "Rosette" is another jointing method which is particularly suited to the prefabrication of light gauge steel frames (Makelainen and Kesti, 1999). A Rosette joint is made between a preformed hole in one of the parts to be joined and a collared hole in the other. The parts are snapped together and then a special hydraulic tool is used pull back the collar and crimp it over the non-collared part of the connection, as shown in Fig. 3. A Rosette typically has a nominal diameter of 20 mm and a strength several times that of a press joint or conventional mechanical connection such as a screw or blind rivet.

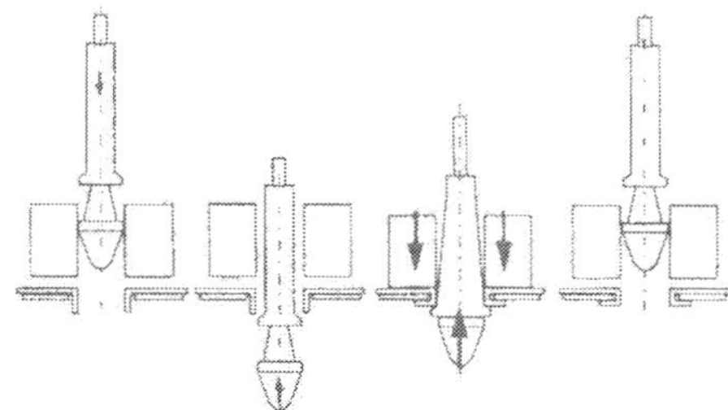

Figure 3. The "Rosette" jointing process (Makelainen and Kesti, 1999)

7.4.4 Fire Safety

In conventional multi-storey construction, fire has been perceived as the "Achilles heel" of steel framed solutions and has tended to direct designers towards the use of reinforced concrete. This has been the driving force for a huge development in the fire engineering design of steel structures and a search for solutions which reduce, or even eliminate expensive fire protection from the economic equation. As a result of this research, Eurocode 3 (EC3) includes simple models for fire engineering design based on first-order theory.

With the important exception of composite floors, the same is not true for light gauge steel construction where there has been almost no comparable research. This means that the EC3 calculation models have not been extended to include cold-formed sections. There may be a number of reasons for this:

- Many of the considerations are similar to those for hot-rolled construction so that separate research into the performance of cold-formed sections in fire may be considered to be less necessary. For example, the use of cold-formed sections is not relevant to the risk of a fire starting, to the spread of smoke or the design of fire-safe escape routes.

- Cold-formed sections are generally used in low rise construction where the fire resistance requirements may be less onerous.

- Development of cold-formed steel framing solutions has been slow, particularly in Europe, so that the necessary research may not have been considered to be cost-effective.

Nevertheless, the use of cold-formed sections is relevant to the risk of structural collapse or the spread of fire to other compartments of the building. There are, therefore, some special consideration with cold formed steel framing which are associated with the thin material and the rather different fire protection arrangements found with this form of construction:

- Structural collapse in a fire occurs when the influence of the rise in temperature reduces the load bearing capacity of the structure to less than the actual load. All structural materials decrease in strength with a strong rise in temperature. In the case of unprotected steel sections, the high thermal conductivity and the relatively small thermal capacity result in a rapid temperature increase of the whole section. This is particularly true for cold-formed sections where the thin material means that the thermal capacity is even more limited.

- For hot-rolled sections, the critical steel temperature, at which structural failure may be anticipated, is in the region 450 - 700^{0}C, depending on the structural system and the load level in the vicinity of the fire. In the absence of any calculation model, a conservative approach for cold-formed sections is to design the fire protection system to achieve a critical temperature of 350^{0}C (EC3, 1995). Some recent fire tests on rectangular hollow sections (Ala-Outinen T and Myllymaki J, 1995) with class 4 cross sections show a critical temperature which is well above 350^{0}C. The same paper also proposes a calculation model in which the effective width of the slender elements of the cross-section is determined using the same formula as at ambient temperature but with the yield strength and modulus of elasticity reduced according to the curves given in EC3. This model gives results which are safe when compared with the test results.

- However, as shown later, with mono-symmetrical cold-formed steel studs, the temperature distribution is likely to be significantly non-uniform. This renders any conventional local buckling theory inappropriate so that special design procedures need to be developed. Evidently, this is a subject worthy of further study and this research is currently in progress in the author's laboratories at the University of Manchester

- Even if the load bearing structure is protected in order to ensure that no structural collapse occurs in a fire, there may still be a risk of fire spread to adjacent compartments through the walls or floors. Light gauge steel floors or steel stud walls must therefore, retain their integrity in a fire for an adequate time and have sufficient thermal insulation to ensure that there is not excessive heat transfer through them.

The fire resistance of structural elements is usually expressed in terms of the resistance time with reference to a standard furnace test which follows a specified time-temperature curve. This curve is according to ISO 834 and national specifications in Europe are generally in accordance with this. A harmonised European classification system has now been adopted which gives separate times for resistance (load bearing capacity), integrity (separating capability) and insulation. Evidently, the required fire resistance time will differ for different types of buildings. It may also differ from country to country for otherwise similar buildings. For residential and office buildings of one or two stories, a normal requirement would be one hour fire resistance for load-bearing elements (R60) and one hour fire resistance for separating elements (EI60).

In conventional hot-rolled construction, this fire resistance may be provided by protection applied to the individual members in the form of boards or sprayed on materials which delay the temperature increase in the steel by providing thermal insulation. However, cold-formed steel members in residential buildings are more often part of a wall or floor structure such that they do not require specially applied fire protection of this sort. Walls and floors are usually built up of layers of materials such as mineral wool and gypsum board in order to meet the requirements of sound and thermal insulation and these materials generally also have favourable properties of fire insulation. Thus, these elements of the construction can be built up in such a way that they meet all of the performance requirements with regard to resistance to fire, sound and thermal loss in a consistent manner. There are some practical points to note here:

- Some fire resistance materials such as gypsum board contain significant quantities of entrapped moisture. Gypsum contains about 20% of crystalline water and, during the time that it takes to evaporate this water, the temperature remains constant at about 100^0C. Depending on the number and thickness of the gypsum boards, this effect can delay the rise in the steel temperature by between 15 and 40 minutes.
- Ordinary gypsum boards start to disintegrate rapidly after the evaporation of the entrapped moisture whereas special gypsum fire boards, which are usually reinforced with glass fibres, stay intact much longer.
- Mineral wool insulation based on stone is preferable to mineral wool based on glass because the temperature softening point is higher in the former.

Node	30 minutes	60 minutes
1	416	740
2	343	638
3	220	478
4	137	338
5	84	215
6	49	107
7	35	92

Predicted temperatures (^{0}C)

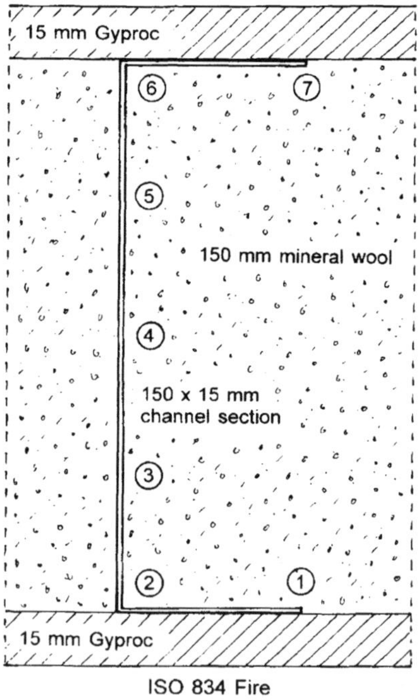

Figure 4. Temperature distribution in a steel stud in lightweight wall construction

The fire performance of such walls and floors generally has to be determined by test. However, there is increasing interest in the use of numerical methods in order to reduce and eventually eliminate the need for expensive testing. TASEF (Wickstrom, 1990) is a well-known computer program which can successfully predict the temperature distribution in assemblies of steel members and insulation products. Fig. 4 shows the results of such a calculation and gives the nodal temperatures after 30 and 60 minutes exposure to the ISO fire for a steel channel within a lightweight wall. It is important to note that the temperature distribution throughout the section is not constant. This immediately requires a more sophisticated analysis than has been applied to hot-rolled sections, particularly with regard to the treatment of both local and global buckling.

Computer programs such as TASEF successfully model the variation of the thermal properties with temperature and the evaporation of entrapped moisture as discussed above. Future developments will include developing more sophisticated numerical models which couple the thermal and structural behaviour in a single finite element analysis.

7.4.5 Durability

With good detailing, the durability of a cold-formed steel-framed structure should equal that of conventional construction. There are two related aspects to ensuring good durability:

- avoid the accumulation of moisture in the external envelope.
- avoid bringing incompatible materials into contact with each other.

These are related because most forms of material incompatibility are aggravated in the presence of moisture. It is first of all necessary to ensure that the cladding is sufficiently impermeable and that it is detailed in such a way that rain water does not penetrate to the structural frame of the building. It would be unwise to assume that all materials and joints will maintain their integrity for ever so that it is also necessary to ensure that any water that does accumulate within the building envelope can escape and that there are no water traps. It is also necessary to ensure that the external envelope performs thermally in such a way that condensation of moisture from humid air is not possible within the building envelope. These are matters of elementary building physics.

The first form of material incompatibility to avoid is bimetallic contact. This is only likely if the cladding or fasteners are of a different material to the main structure. Other situations to avoid are:-

- *With cementitious materials*: Galvanised steel reacts with fresh cementitious materials with the formation of hydrogen bubbles on the surface of the zinc which reduce the bond with the steel. This reaction can be controlled by the addition of soluble chromates to the cementitious material but a more reliable solution is to chromate the surface of the galvanised steel.
- *With gypsum plaster*: Moist gypsum plaster attacks galvanised steel both in the fresh condition and if it dries out and subsequently becomes wet. The G275 coating of normally galvanised steel will prevent corrosion of the underlying metal during the drying out period and, provided that the plaster then remains dry, no further corrosion will take place.
- *With timber*: Wood is a naturally corrosive material which can be made more corrosive by the protective treatments given to it. Some species, such as oak, can also give off acidic vapour which, in a poorly-ventilated space, can cause corrosion of nearby metals

which are not actually in contact. However, correct seasoning of the timber significantly reduces the risk. Generally, there is less risk with soft woods than with hard woods. Corrosion of metals in contact with timber is a function of the moisture content of the timber. Wood is not normally used where the moisture content may remain above 18% for significant periods. Above 22% moisture content, there is a risk of degradation of the timber and it is also under these conditions that metallic components would corrode in timber. It follows that galvanised steel components in contact with some treated timbers may be at risk of corrosion. Where the steel is coated with a material which is, itself, impervious to attack from the timber and its preservatives, there is unlikely to be any degradation of the steel.

7.4.6 Acoustics

Sound transmission and attenuation is a complex subject. It is also important because the occupants of buildings are rightly sensitive to noise intrusion from other occupants of the same building. Space precludes a full discussion of this topic in the present paper and a few simple principles will have to suffice.

The aim is to construct walls and floors in such a way that they provide the maximum reduction in sound transmission measured in decibels (dB). Requirements are laid down by a number of countries in terms of the required airborne sound reduction index R_w' between the rooms of a completed building. These are also indicative of the requirements to avoid intrusive noise from outside. The following are typical:

between apartments in residential buildings;	50 - 55 dB
between guest rooms in hotels;	48 - 52 dB
between classrooms in schools;	40 - 45 dB

As a guide, the following list indicates the consequences of different values of R_w' on the audibility of the human voice from one room to another:

loud speech cannot be heard through a wall;	55 dB
normal conversation cannot be heard through a wall;	45 dB
normal conversation can just be heard but there is little impact on working in the other room;	35 dB
below this value, the wall is little more than a visual screen;	30 dB

Structures can be categorized on the basis of their acoustical behaviour:

- single leaf (massive) structures
- double leaf structures
- multi-leaf structures
- single leaf structure supplemented by additional cladding

7.4.6.1 Single leaf structures

A heavy structure provides better sound insulation than a light one (the mass law) so that the basic rule for single leaf structures is "the heavier the better". However, the sound transmission loss grows only slowly with increasing mass to the extent that doubling the mass gives only a 6 dB increase. It follows that single leaf structures are not very effective in lightweight construction and that, for walls in particular, double leaf solutions are to be preferred.

7.4.6.2 Double leaf structures

Double leaf walls give better insulation than single leaf walls, especially for high frequency sounds. Ideally, a double leaf structure should be made up of two separate sheets with the space between them filled with a soft sound-absorbing material such as mineral wool. These sheets should not, in any way, come into contact with each other. The wider the cavity between the sheets and the greater the weight of the sheets, the better the sound transmission loss. Fig. 5 illustrates the airborne sound reduction achieved by single and double leaf structures with steel sheets. In practice, the two leafs of a double leaf wall often have to come into contact with each other. The consequence of this is a significant weakening of the sound transmission loss.

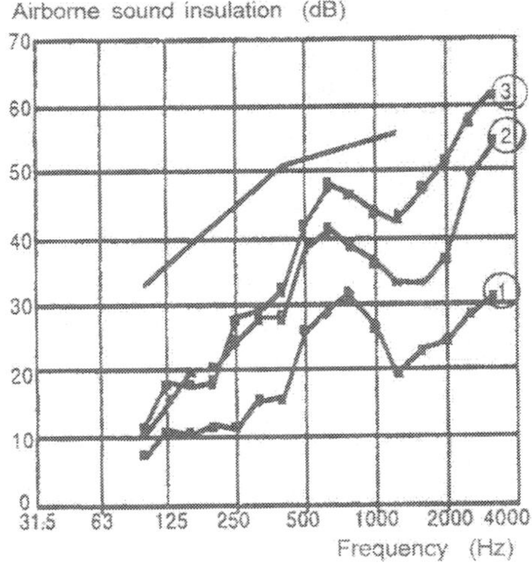

1: profiled steel sheet, thickness 0.8 mm

2: profiled steel sheets, 2 × 0.8 mm, 120 mm air space

3: profiled steel sheets, 2 × 0.8 mm, 120 mm mineral wool in cavity

Figure 5. Airborne sound insulation provided by profiled steel sheets

7.4.6.3 Multi-leaf walls

Walls with three or more leafs work in the same way as double-leaf structures. The system may be considered as several masses with air "springs" between them and it has as many resonant frequencies as there are air spaces. The lowest frequency is at approximately the same point as with a double leaf wall of the same weight and thickness and the others are higher than this. As a consequence, the sound transmission loss of a multi-leaf wall is usually weaker than that of the corresponding double-leaf structure.

7.4.7 Thermal insulation, air and moisture permeability

In the design of walls and roofs of lightweight construction, the related subjects of thermal insulation, air and moisture permeability are paramount. The need to achieve an adequate level of thermal insulation is now well understood and the statutory requirements seem to become more onerous year by year. A fundamental feature is the need to avoid "thermal bridging". It is less well understood that this theoretical thermal performance can be destroyed by air leakage, particularly at the eaves of a building. Avoiding this is a matter of detailing. It is also well understood that it is necessary to avoid condensation of moisture within the structure and that a primary requirement is an appropriately placed vapour barrier. Control of moisture is then also largely a matter of good detailing.

The fundamentals of typical "warm frame" wall construction are shown in Fig. 6. For most climates, the first essential is a high level of thermal insulation and this generally involves the provision of a suitable thickness of insulation material as well as a cavity. The insulation material is generally mineral wool although a wide variety of suitable alternatives is available. It is also essential to avoid thermal bridges and moisture penetration.

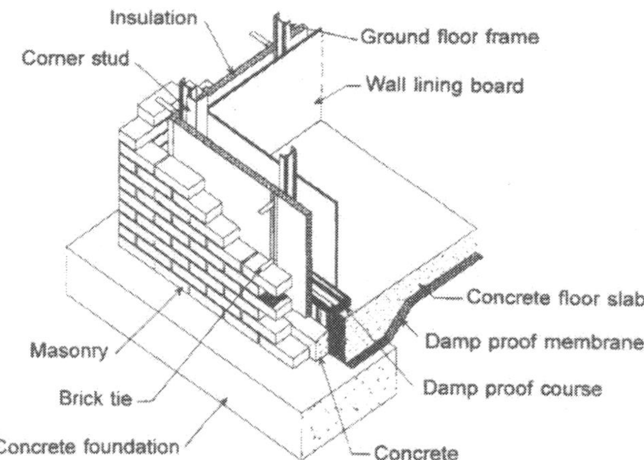

Figure 6. Typical wall construction for domestic buildings (Grubb and Lawson, 1997)

Further details of wall construction, including connection details at an intermediate floor and at the roof, are shown in Fig. 7.

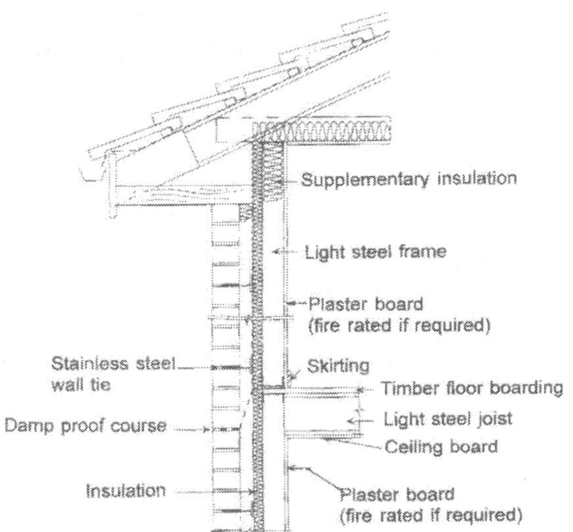

Figure 7. Typical wall details at the eaves and intermediate floor (Grubb and Lawson, 1997)

7.4.8 The role of the Structural Engineer in steel frame house construction

It is clear from all of the foregoing that the structural engineering aspects of steel frame house construction are by no means trivial and that, perhaps, the development to date has suffered from a lack of imaginative engineering. It is equally clear that, in the development of this form of construction in Europe, Structural Engineers will have a significant role to play. However, it must also be appreciated that too heavy a dependence on detailed structural design represents a considerable disincentive towards this form of construction. To this end, the major countries involved in this form of construction have provided detailed design manuals giving considerable amounts of information and design aids (eg AISI, undated; CSSBI, 1991).

With more conventional construction materials such as brickwork and timber, prescriptive methods, involving the use of design charts and tables, have been developed and incorporated into codes of practice with the result that the services of an Engineer are not usually considered to be necessary for the design of a typical house. If steel framed construction is to flourish in the UK and Europe, a similar approach needs to be developed, at least for simple, relatively standard designs.

Recognising this, the US Department of Commerce has recently published the First Edition of the "Prescriptive Method for Residential Steel Framing" which addresses this problem. The Second Edition of this document (AISI, 1997), together with a detailed Commentary, is included in the Residential Steel Framing Manual (AISI, undated).| Evidently, something similar is required for European practice.

7.4.9 Conclusions

This chapter has reviewed the current state of the art with regard to the steel framed construction of houses and similar low rise construction. The prospects are exciting in that there is scope for a dramatic rise in this form of construction and the industry is responding by offering appropriate solutions. However, at the same time, the situation is disappointing in that, with the possible exception of cassette construction, no new architecture or technology appears to be emerging.

Sufficient has been written to make it clear that the design of the steel frame should be inseparable from the choice of cladding and lining materials and these, in turn, are strongly influenced by the architectural design of the elevation. Existing new cladding materials are appearing on the market yet we still have the situation (notably in the UK) where architects are designing conventional elevations for brick construction and asking engineers to hide steel frames behind them and line them with plasterboard. The result is a house which is completely indistinguishable from one in conventional masonry construction.

The author can fully appreciate the reasons why this situation has come about. They lie in the conservative nature of "middle England" and the desire of developers to minimise risk. At the same time it is not easy to see how this bind may be broken and perhaps here the structural steel industry may have to take the initiative. What is required is a distinctively steel house in which the architecture and cladding are chosen to be fully compatible with a steel frame. The different professions involved will not arrive at a satisfactory solution independently. They need to work together starting with a blank sheet of paper. Then, perhaps, we may arrive at the economic and attractive house designs that are appropriate for the 21st century.

References

AISI (1996) *American Iron and Steel Institute.* Specification for the design of cold-formed structural members.

AISI (1997) Prescriptive method for residential steel framing; Second edition, Prepared by NAHB Research Center, Upper Marlboro, MD for the American Iron and Steel Institute, August, 93pp.

AISI (undated) "Residential Steel Framing Manual". *American Iron and Steel Institute.* Washington DC.

Ala-Outinen T and Myllymaki J. (1995) "The local buckling of RHS members at elevated temperatures". VTT Research Notes 1672, *Technical Research Centre of Finland.*

Baehre, R. and Bucca, J. (1986) "Die wirksame Briete des Zuggurtes von biegebeanspruchten Kassetten" (Effective width of the tension flange of cassettes in bending), *Stahlbau* 9, 276-285.

Baehre, R. (1987) "Zur Shubfeldwirkung und-bemessung von Kassettenkonstructionen" (On the behaviour and design of cassette assemblies in shear*)* *Stahlbau* 7, 197-202.

Baehre, R., Buca, J. and Egner, R. (1990) "Emfehlungen zur Bemessung von Kassettenprofilen" (Recommendations for the design of Cassettes), *R. Schardt Festschrift*, University of Darmstadt.

BSI (1996) BS 5950: Part 9. Structural use of steelwork in building: Code of practice for stressed skin design, *British Standards Institution.*

CSSBI (1987) *Canadian Sheet Steel Building Institute*. Lightweight steel framing manual. August, 23pp.

CSSBI (1991) "Lightweight Steel Framing Design Manual". *Canadian Sheet Steel Building Institute*. July. 102pp.

Davies J M and Fisher J (1979). The diaphragm action of composite slabs. *Proceedings of the Institution of Civil Engineers: Part 2*: Vol. 67, December, 891-906.

Davies J M and Bryan E R (1981) Manual of stressed skin diaphragm design, *Granada*.

Davies J M (1986) A general solution for the shear flexibility of profiled sheets, I: Development and verification of the method and II: Applications of the method, *Thin Walled Structures*, Vol. 4, No. 1, 41-68 and Vol. 4, No. 2, 151-161.

Davies J M and Fisher J (1987) "End failures in stressed skin diaphragms", *Proceedings of the Institution of Civil Engineers, Part 2,* Vol. 83, March, 275-293.

Davies, J. M. and Hakmi, M. R. (1991) "Post-buckling behaviour of foam-filled thin-walled beams", *J. Constructional Steel Research*, Vol. 20, 75-83.

Davies, J.M., Leach, P. and Heinz, D.(1994) "Second-order Generalised Beam Theory", *J. Constructional Steel Research*, Vol. 31, Nos. 2-3, 221-241.

Davies, J.M., Leach, P. and Kelo, E. (1995) "The use of light gauge steel in low and medium rise modular buildings*", 3rd Int. Conf. on Steel and Aluminium Structures, ICSAS '95*, Istanbul, 24-26 May.

Davies, R, Pedreschi, R and Sinha, B. P. (1996) "The shear behaviour of press-joining in cold-formed steel structures", *Thin-Walled Structures*, Vol. 25, No. 3, 153-170.

Davies, J. M. and Dewhurst, D. W. (1997) "The shear behaviour of thin-walled cassette sections infilled by rigid insulation", *Int. Conf. on Experimental Model Research and Testing of Thin-Walled Structures*, Academy of Sciences of the Czech Republic, Prague, Sept. 209-216.

Davies J M (1999) The use of cold-formed steel in residential and office buildings, *Int. Conf. on Steel and Composite Structures*, 24-25 February, 12.1-12.12.

Davies J M and Lawson R M (1999) "Stressed skin design of modern steel roofs", T*he Structural Engineer*, Vol. 77, No. 21, November.

EC3 (1995) "Eurocode 3: Design of Steel Structures, Part 1.2: Structural Fire Design". CEN ENV 1993-1-2.

EC3 (1996) Eurocode 3: Design of Steel Structures: Part 1.3: General Rules: Supplementary rules for cold formed thin gauge members and sheeting. *European Committee for Standardisation*, ENV 1993-1-3.

ECCS (1995) European recommendations for the application of metal sheeting acting as a diaphragm, *European Convention for Constructional Steelwork*, Publication No. 88.

Engel P (1998) "Recent developments in dry floors using light gauge steel components", Paper No. 62, *2nd World Conf. in Steel construction*, San Sebastien, 11-13 May.

Grubb P J and Lawson R M (1997) Building design using cold formed steel sections: Construction detailing and practice. SCI Publication P165. *Steel Construction Institute,* 119pp.

Griffiths D R and Wickens H G (1995) "Final Report on diaphragm tests on braced steel framed wall panels", *Dept. of Civil Engineering, University of Surrey*, July.

Helenius A (2000) Shear strength of clinched connections in light gauge steel, VTT Research Notes 2029, *Technical Research Centre of Finland.*

Hoglund T (1998) Design of light-gauge studs with perforated web. *Nordic Steel Construction Conference,* Bergen, Norway, Sept. 14-16, 811-824.

Kesti J and Makelainen P (1998) Design of gypsum-sheathed perforated steel wall studs. *2nd World Conference on Steel in Construction*, San Sebastian, 11-13 May.

König, J. (1978) "Transversally loaded thin-walled C-shaped panels with intermediate stiffeners", *Swedish Council for Building Research*, Document D7.

Makelainen P and Kesti J (1999). Advanced method for lightweight steel joining. *Journal of Constructional Steel Research*, 49, pp 107-116.

Miller T H and Pekoz T (1994) Unstiffened strip approach for perforated wall studs. *Journal of Structural Engineering*. Vol. 120. No. 2. February.

Nyberg, G. (1976) "Diaphragm action of assembled C-shaped panels", *Swedish Council for Building Research*, Document D9.

Serrette, R. L. and Pekoz, T. (1995) "Distortional buckling of thin-walled beams/panels. I: Theory and II: Design methods", *Journal of Structural Engineering*, Vol. 121, No 4, April 757-766 and 767-776.

Thomasson, P.O. (1978) "Thin-walled C-shaped panels in axial compression" *Swedish Council for Building Research*, Document D1.

Wickstrom S (1990) "TASEF - Temperature Analysis of Structures Exposed to Fire". *Swedish National Testing and Research Institute*. Report 1990:05.

Yu W-W (2000) Cold formed steel design, Third Edition, *John Wiley & Sons Inc*, New York, 357-359 and 547-567.

Chapter 8: Industrial and Non-residential Buildings

D. Dubina

Department of Steel Structures and Structural Mechanics, Civil Engineering and Architecture Faculty, "Politehnica" University of Timisoara, Romania
E-mail: dubina@constructii.west.ro

8.1 Generals

Statistics of European Convention for Constructional Steelwork - ECCS - shows that actual annual production of constructional steelwork is of about 6.5-7.0 millions tones. A vast majority of this amount is fabricated from conventional hot-rolled steel. However, there are applications for which the cold-formed steel is both technologically more competitive and costly effective if compared with conventional technology. There are significant advantages of using cold-formed steel sections, in regard with building erection time, aesthetics, prefabrication, quality assessment, flexibility of adaptation, recyclability, lightness, low maintenance and environmental friendliness. Based on so called ***'Mixed Building Technology - MBT'***, cold-formed steel can be used in association or combination with timber, masonry, concrete, glass, gypsum, insulation materials, which can result in optimised and higher performance buildings.

The most common applications of cold-formed steel sections in industrial and non-industrial buildings are summarised on the following:

- Roof and wall members (purlins and rails)

Traditionally, a major use of cold-formed steel has been, and continues to be, as purlins and side rails to support roofing and wall cladding in industrial type buildings. They are generally based on Z sections (lipped channels or variants being also possible) which facilitate incorporation of sleeves and overlaps to improve the efficiency on the member in multi-span applications.

- Steel framing

An increasing market for cold-formed steel sections is *'on site'* assembled frames and panels for walls and roofs, and stand-alone buildings. This approach can be used in light industrial and conventional (e.g. non-residential) buildings and in mezzanine floors, but also when restructuring existing buildings by additional storey or penthouse structures.

Different structural systems can be used for such kind of applications, made by:

- built-up sections obtained from Swagebeams, Walleybeams, Sigma sections (Frisomat System) strong stiffened channels (Conti System), back-to-back spaced lipped channels, etc;
- single open sections, for small spans only, as haunched press backed Z sections (Galcorom System), lipped channels (Lindab Kft. System);
- RHS sections.

Combined systems such as dual systems of frames with wall stud based shear walls or truss frames are also used. Usually, bolted connections are used, but for small size buildings also screwed connections can be applied.

- Floor joists

Single or built-up floor beams by C and Z (or variants) sections are possible.

- Truss

There are a number of European manufacturers of lattice girders and truss systems using cold-formed steel sections like SADEF (Belgian), MEETSEC (UK), LINDAB (Sweden, Romania, etc.), and others.

Single and built-up sections are used as members. Bolted, screwed and, recently, press joining and rosette techniques are used for connections.

- Roof decking and wall cladding

Sheeting, sandwich panels and cassettes are available for these applications. In case of sheeting, diaphragm effect (stress skin action) can be considered if appropriate fastening technique is used.

- Floor decking

Floor decking is used for dry and composite steel and concrete slabs in multi-storey frame buildings. Also, in this case, the diaphragm effect is very important, especially for buildings in seismic zones. The market for this application continuously increases. Both conventional and slim floor systems are now available on the market.

- Plane and arched self-bearing sheeting for buildings envelops

Deep corrugation sheeting do not need for any supporting system (purlins and rails).

- Frameless buildings

This is an extension of previous applications to the whole building. Frameless buildings are made by steel folded plates, barrel vaults, truncated pyramid roofs and hyperbolic paraboloids, have been developed for different applications.

Corrugated sheeting is generally used for such kind of applications. They are called 'frameless buildings', because they are self-bearing systems (e.g. without beams or framing supporting structures) which also rely partly on 'stress skin' action. Corrugated sheeting can be also used to build beams or even portal frames.

In this section, steel framing and truss systems will be examined only. On this purpose, selected study cases, based on design applications by the author will be further presented. After, some particular features of these structures will be detailed, i.e.:

- behaviour of built-up back-to-back channel members with bolted stitches;
- structural properties of bolted connections in cold-formed steel framing.

At the end, strength and ductility of wall stud based shear panels will be examined. Their performances are important when they are integrated in dual framing structures, in which they have to carry horizontal static and dynamic loading from wind and seismic actions, respectively.

8.2 Case studies. Structural performances

This section presents three types of cold-formed steel buildings. The buildings are all located in heavy snow and seismic zones and are fully made by cold-formed steel sections.

The framing structures are fabricated by single C and built-up spaced back-to-back C sections. Only bolted connections are used. Prefabricated units can be also obtained. The diaphragm effect of roof sheeting panels and some vertical braces are used to resist against horizontal actions loading from wind and seismic actions.

The structures are different as destination and conception, i.e.:

- Wall Stud Modular System (WSMS) - bungalow type buildings, are used as shops, offices, or industrial purposes, for housing or school facilities;
- Hall Type Framed Structure (HTFS) – are used as small and medium size single storey industrial buildings and storehouses;
- Penthouse Framed Structure (PFS) – for refurbishment and restructuring by vertical addition of new storey the existing buildings; they can be used as industrial buildings, offices, education facilities, shops etc.

General design principles of cold-formed structures and building technologies are presented together with some specific design detailing, accompanied with examples of application.

Wall Stud Modular System (WSMS). The main idea in designing the WSMS (Dubina et al., 2000, 2001) was to combine the "wall stud" system, used for steel-framed housing, with a "light roof" solution, used for industrial hall type buildings. Figure 1 shows the basic components of the system. The 3D model comprises main frames, with built-up sections in columns and trusses (braced Vierendell System), the wall-stud panels and the roof structure with purlins and sheeting panels (diaphragm effect is considered), able to carry both vertical and horizontal loads. The "wall stud" panels are adaptable, according to the shape and dimensions of openings. All structural and non-structural components are made of cold-formed steel LINDAB CONSTRULINE sections. The structure is designed by modules, to be installed bay-by-bay, using the so-called "stick" technique, easier to transport and to build pre-erected units (panels, trusses) on the site. Prefabrication in modules is also possible. The façade as well as the choice of foundation are decided by the customer in co-operation with the retailer and according to technical specifications given by designer.

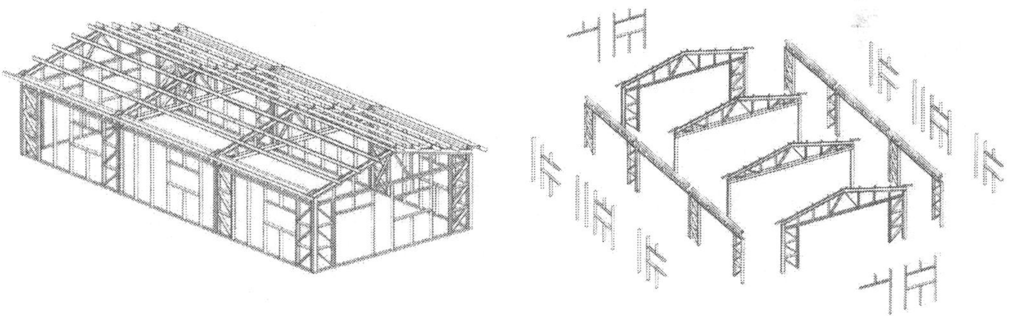

Figure 1. Main frame structure and basic components of a WSMS

Table 1 contains the characteristics of snow and seismic load data and main dimensions of the actual WSMS version, while Figure 2 shows the theoretical (computer output) weight performances (steel consumption for the main frame structure, e.g. without purlins and adaptable wall panels) of the system (Dubina et al., 2001). Table 2 displays the main dimensions, the location and the real weight (design specification for main frame structure) of some WSMS units erected in Romania in 2000 and 2001, while Figures 3 to 6 are showing some representative images for these buildings.

Table 1. Characteristics of snow and seismic actions and main dimensions of the actual WSMS

Region	Main dimensions			Design snow load g_z (kN/m^2)	Seismic parameters*
	Span S (m)	Bay B (m)	Eaves Height H (m)		
Bucharest	8-12	5 (6)	3 (3.5)	3.12	$\gamma_I = 1$ $\alpha = 0.2$ ($T_c = 1.5$s); $q = 1$
Timisoara	8-12	5 (6)	3 (3.5)	1.8	$\gamma_I = 1$; $\alpha = 0.2$ ($T_c = 1.0$s); $q = 1$

*) The seismic parameters were determined according to EUROCODE 8 (ENV 1998)

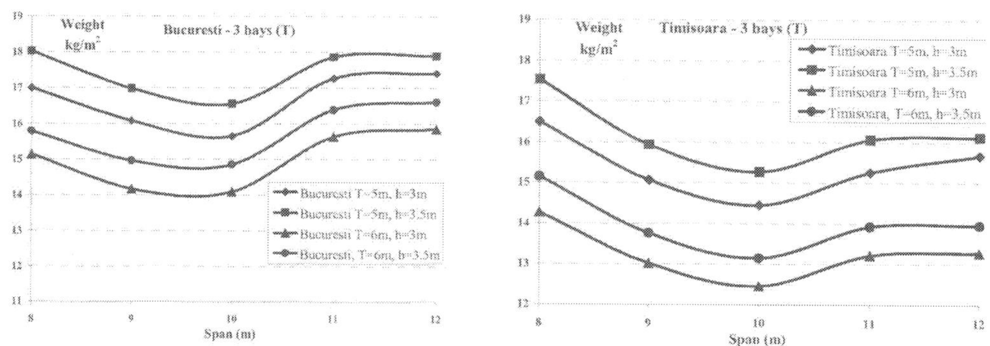

Figure 2. Weight performances of the system

Table 2. Main dimensions, location and the weight of some WSMS units

Name of structure	Dimensions S×(nB) ×H (m)	Location	Weight (kg/m^2)
ROMRECYCLING Bucharest, Office building	10×(2×6) ×3	Bucharest	20
JASMIN Satu-Mare, Storehouse for Dr. Oetker	10×(3×6) ×3	Satu-Mare	21
AQUABIS Bistrita-Nasaud, Storehouse	10×(2×6) ×3	Bistrita-Nasaud	22
FLAMINGO Bucharest, Storehouse	10×(3×6) ×3	Bucharest	20
TERANITA Giurgiu, Storehouse	10×(4×6) ×3	Giurgiu	21
ALCEDO Timisoara, Storehouse	9.65×16.75×3.9	Timisoara	24
BIOLACT Oradea, Storehouse	10×(7×5.4) ×3	Oradea	23

Figure 3. Main frame structure of BIOLACT building

Figure 4. ROMRECYCLING structure during erection

Figure 5. JASMIN building

Figure 6. ALCEDO building

Hall Type Modular System (HTMS)

For eaves height greater than 3.5 m the WSMS (Dubina et al., 2000, 2001) becomes inefficient to be used. For such situations, which may appear quite often in practice, a cold-formed steel structural system based on pitched roof portal frames made by built-up back-to-back "C" sections connected with stitches of Johnston type (Johnston, 1971) was developed. The main components of this system are shown in Figure 7.

Two different buildings, designed in 1999-2000, using Johnston built-up cold-formed sections are presented in Table 3. Due to the fact the cold-formed steel sections are of class 4, the structures are considered to work in the elastic range, being non-dissipative, and correspondingly, the q coefficient was taken equal to 1.

The structures are designed accounting for diaphragm action of sheeting (Dubina et al., 2000) (ECCS, 1995). In Figure 8 are shown the steel consumption curves for some usual dimensions of a building located in Bucharest.

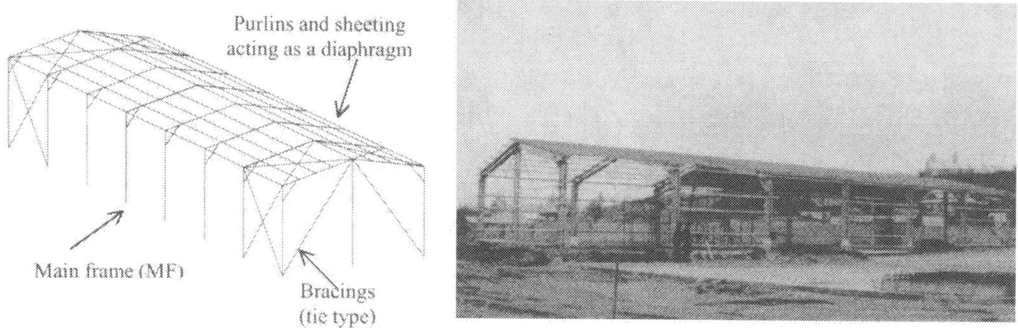

Figure 7. Main frame structure and basic components of a HTMS

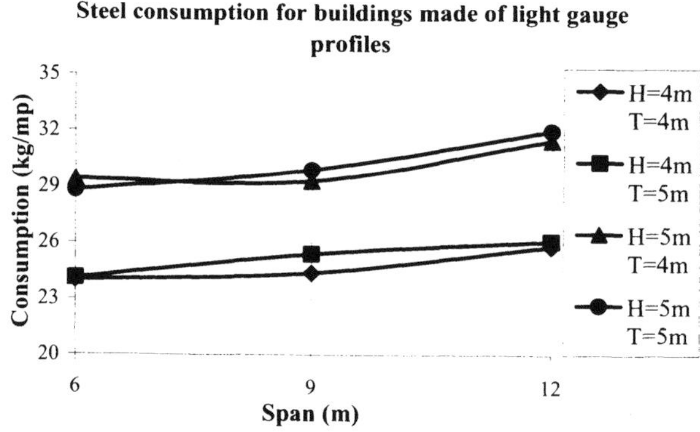

Figure 8. Steel consumption curves (Bucharest)

Two relevant examples of application of HTMS system are the BRAU-Union Brewery thermal power station building located in Arad, in the west part of Romania, near to Timisoara and POPET Hall in Vaslui, located in the east part of Romania (see Table 3). Figures 9 and 10 show the framing structure and accomplished building of BRAU-Union Hall, respectively.

Table 3. Main dimensions, location and the weight of some HTMS units

No	Name	Location	Type of building	Main dimensions			Weight
				L [m]	B [m]	H [m]	(kg/m^2)
1	BRAU-Union Hall	Arad	Single story house	12	42	7.00	31
2	POPET Hall	Vaslui	Single story house	12	36	4.28	21

Figure 9. HTMS structure in Arad City

Figure 10. BRAU-Union Hall after completion

Penthouse Framed Structures (PFS). Restructuring by vertical addition consists of adding one or more stories above the existing structure, resulting in an increase of overall volume of the building.

Depending the size and height of the new addition masses, it is necessary to re-analyse the load-bearing capacity of the original structure in order to decide whether or not to take consolidation measures. In seismic zones this problem can be very serious, and traditional building techniques cannot be always used for this type of restructuring. The necessity to minimise the weight of the new storey structure added above, makes cold-formed steel sections the most suitable solution.

Cold-formed steel structures made by built-up members of Johnston type (Johnston, 1971), by LINDAB C sections, and special bolted connections are very efficient for penthouses above the existing masonry or RC buildings. The new floor, because the existing terrace (RC slab or timber structure) usually has not enough load-bearing capacity to carry the additional structure, with its dead and life loads, is made by a grid of steel castellated beams.

Two application examples of this constructional steelwork solution focussing mainly the cold-formed steel structure of penthouse are shown further (Dubina et al., 2000, 2001).

Table 4. Thin-walled cold-formed structures with Johnston type cross-sections for PFS

No.	Name	Location	Type of building	Main dimensions		
				L [m]	B [m]	H [m]
1	ALCATEL-DATATIM Hall	Timişoara	Penthouse	24	24	4.00
2	CANTEMIR Hall	Tg. Mureş	Penthouse	18	30	3.55

Penthouse structure for Alcatel-Datatim Company in Timisoara. The structure described in the paper represents an extension of the production capacity belonging to ALCATEL-DATATIM Company of Timisoara (Dubina et al., 2000).

By project initial data, an extension of the existing production space was required by:

a) erection of a ground floor structure over the existing courtyard;

b) erection of a penthouse superstructure over the buildings;

c) erection of a staircase in the space of the access gallery, to provide access from downstairs to the first floor level;

d) extension of the social outbuilding also by penthouse structure.

The main characteristics of penthouse component buildings are given in Table 5.

Due to the fact Timisoara is a seismic territory the penthouse structure had to be light enough in order to avoid a heavy and expensive strengthening of existing RC structure, not properly designed according to seismic design criteria. However, the RC structure was supplementary braced.

A thin-walled cold-formed structure connected exclusively by bolts was for the penthouse superstructure. The roofing and cladding panels are of LINDAB trapezoidal steel sheet, with suitable thermal insulation. As interface structure, a metallic grid built of castellated steel beams interconnected by HSFG bolts and connected to the concrete structure by cruciform steel elements was provided. The floor of the penthouse is a composite steel-concrete deck (Dubina et al., 2001).

Table 5. Main characteristics of penthouse component buildings

Name	Destination	Deck Level [m]	Eaves Height [m]	Built Area [m^2]	Built Volume [m^3]
Penthouse over buildings	Production	+5,70	4,00	1066	4264
Social outbuilding	Social	+5,00	3,00	85	255
Greenhouse	Social	±0,00	8,00	90	720
Stair case + Stairs	Access to first floor	±0,00	9,00	36	320

In order to avoid to overload the existing structure with bending moments from horizontal actions (earthquake and wind), the column supports at both ends are pinned and, correspondingly, the structure had to be properly braced. The diaphragm effect of sheeting in roof structure was also considered. Figure 11 shows the structure during erection. Joint details are given in Figure 12 while Figure 13 shows the building after completion.

The structure was designed on the basis of a 3D static and dynamic analysis. Technical performances proving the efficiency of this structure are shown in Table 6.

A very important advantage of this structural solution, in the particular conditions of narrow building site, was the easy erection. Practically, the structure was raised, stick by stick, on the deck of new floor, and there the trusses and frames were assembled and lifted with a light autocrane.

Figure 11. ALCATEL-DATATIM structure

Figure 12. Eaves connection and ridge connection for ALCATEL Hall

Figure 13. View of the building during erection and after completion

Table 6. Technical performances

Characteristic	Value	Comments
Maximum transversal top drift	19.3 mm	Seismic action
Maximum longitudinal top drift	36.8 mm	
Eigenperiod on transversal direction	0.388 sec	
Eigenperiod on longitudinal direction	0.690 sec	
Steel weight of the skeleton	17 kg/m^2	
Steel weight of purlins and rails	9 kg/m^2	
TOTAL steel weight of structure	26 kg/m^2	

Penthouse structure above a three storey existing masonry infield RC framed building in Tg. Mures. The existing building was an uncompleted three storey masonry infield RC framed construction, built in Tg. Mures, with the initial purpose of some economical activity. That building was bought by "Cantemir" University in the idea to be restructured for teaching activity. An additional storey was estimated to be necessary, but extension with a new masonry storey would be not possible because not enough load-bearing capacity of existing structure was available. Thus, a light gauge steel structure was preferred. Practically, with some small differences, the same type of structure like for ALCATEL-DATATIM has been used. The main cold-formed steel frames are shown in Figure 14, while joint details are presented in Figure 15. Figure 16 shows the penthouse structure with the castellated beam grid interface, after erection, and the whole four storey composite masonry infield RC framed – steel building.

Figure 14. Current and gable frames for Cantemir penthouse

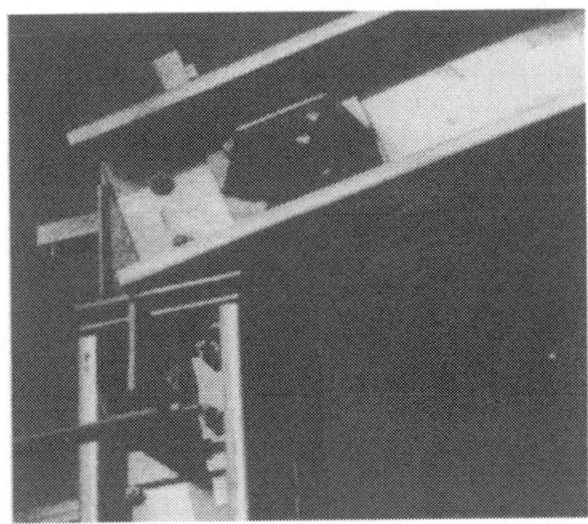

Figure 15. Eaves connection and ridge connection for Cantemir Hall

Figure 16. Penthouse structure erected and whole building

The weight of main framed structure was 18 kg/m^2 only, to which purlins and external walls structure has to be added. Due to the lightness of structure components, no crane was necessary for erection.

8.3 Peculiar design problems

The structural solutions previously presented are characterised by some specific features, i.e.:

- using of built-up members of lipped channel sections connected to each other with stitches;
- tacking the benefit from semi-rigid behaviour of the joint trusses.
- the use of diaphragm effect, particularly to improve the seismic response;

If a static loading design is performed only, the problem can easy be managed on the base of ECCS Recommendations (ECCS, 1995) if a static loading design is performed, only. In case of dynamic design, for seismic resistant buildings, in order to take a benefit from shear performances of roof and /or cladding diaphragms, special analysis is necessary. For the other two problems, also specific design procedures must be used. The following sections focus these specific problems.

8.4 Built-up sections for frames and trusses

Generals. Framing systems presented in the previous section are characterised by using built-up members made of back-to-back C sections connected with bolted C or U stitches. Such type of members can be applied for columns and beams of pitched roof portal frames and also for chords of trusses.

There are no specific recommendations in the European norm EUROCODE 3 (EN1993-1.1, 2001) for this type of battened members. However, the rules for classical battened columns could be used also for this particular case, provided the specific behaviour of thin-walled cold-formed sections is accounted for.

Previous proposal made by Rondal and Niazi (1990, 1993), which is based on formulas given by Johnston (1971) for spaced hot-rolled columns, connected by hinged battens, is the only dedicated design recommendation to calculate the resistance of such cold-formed built-up members in compression.

In case of members in bending there are no particular design rules excepting the fact that the distances between stitches must be appropriately sized in order to comply with built-up section conditions.

Further, the EUROCODE 3 provisions for battened columns will be compared with Rondal and Niazi proposal, via some experimental and numerical results for this particular type of built-up C cold-formed steel sections.

Built-up columns of Johnston type (1971). Johnston (1971) pioneered the study of the behaviour of spaced columns in which the battens are attached to the longitudinal column elements by hinged connections. In this case, battens act as spacers only, with no shear transmitted between longitudinal elements. For this reason, Johnston focused his study on the strengthening effect of end-tie plates on the buckling load (Figure 17). The buckling load of a spaced column with end tie plates is a lower bound to the buckling load of a battened column with low or uncertain moment resistance in the connections between battens and the longitudinal components.

For the hinged end condition, a spaced Johnston column will buckle either in mode A (centre reversal of curvature) or in mode B (semi-fixed shape), as shown in Figure 18.

In mode A of buckling there is no differential change of length between the two members of the battened column. Thus, the two longitudinal column components may buckle in Mode A under identical loads P/2, and the critical load is independent of the ratio I/I_1, where I is the moment of inertia of the whole column and I_1 is the moment of a single section. The stability equation from which the critical load can be obtained in this particular case is:

$$tg\left(\frac{kl}{2}\right) = -ka\,, \qquad \text{with:} \quad k^2 = \frac{P}{2EI_1} \tag{1}$$

In terms of overall length L, the critical load results as:

$$P_{crA} = \frac{C_A EI_1}{L^2} \tag{2}$$

in which C_A is termed buckling coefficient. It is convenient to express the critical load function of the equivalent length coefficient K:

$$P_{crA} = \frac{2\pi^2 EI_1}{(KL)^2} = \frac{C_A EI_1}{L^2} \qquad \text{in which} \quad K = \pi\sqrt{2/C_A} \tag{3}$$

In case of a=0 (no end-tie plates) the equivalent length coefficient becomes K=0.5 and consequently the critical load is:

$$P_{crA} = \frac{8\pi^2 EI_1}{L^2} \tag{4}$$

When the column buckles in mode B, the shortening under column load is greater on the concave side than on the convex one; thus, there is a supplementary internal resisting moment due to direct forces in the components that is added to the bending moments induced in components themselves. The critical load for mode B may be less than these for mode A when the ratio I/I_l is relatively small and a/L is large. Consequently the equation from which the critical load results is:

$$kl\sin kl + \left(\frac{I}{I_1} - 2 - k^2 al\right)(1 - \cos kl) = 0 \tag{5}$$

Further on, the critical load is:

$$P_{crB} = \frac{C_B E I_1}{L^2} \tag{6}$$

For $a=0$, the buckling coefficient C_B may be approximated with a good precision with formula:

$$C = 14.91 \cdot \ln\frac{I}{I_1} + 18 \qquad \text{for } 6 \le \frac{I}{I_1} \le 42 \tag{7}$$

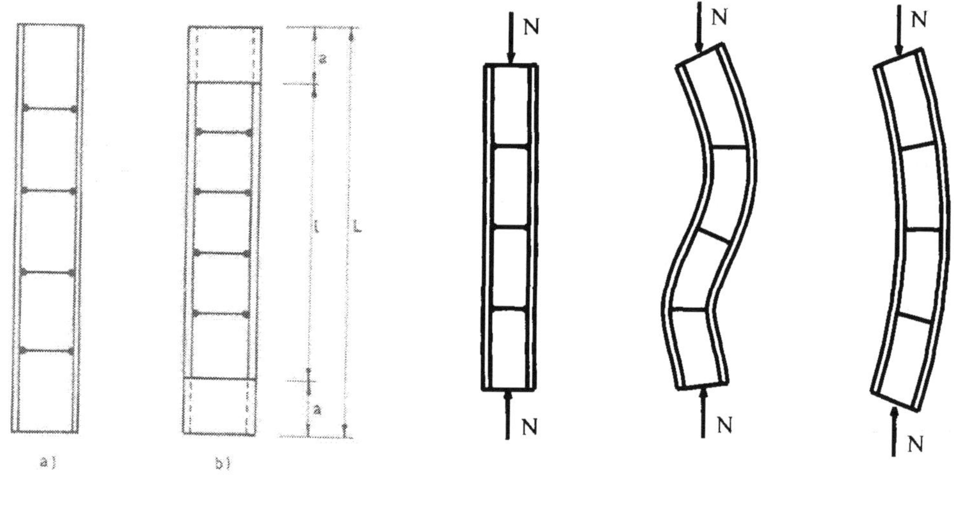

Figure 17. Johnston built-up column: a)without end-tie plates b)with end-tie plates

Figure 18. Model of a column with C stitches

Rondal and Niazi proposal (1990, 1993). The authors calibrated the Johnston formula, previously presented, in order to apply for built-up columns made of C sections connected with bolted C stitches. Indeed, in case of such a built-up section, the junction between component profiles, realised with bolted C stitches is more flexible than conventional battened plates, and may be considered as hinged. Consequently, the critical loads determined by Johnston, without considering the end-ties effect, may be used.

For buckling mode B, Rondal and Niazi proposed a modified equation for the critical load, e.g.:

$$P_{crB} = \frac{\pi^2 E I_{eq}}{L^2} \tag{8}$$

in which I_{eq} is an equivalent second moment of area of the entire cross section of the battened element:

$$I_{eq} = \frac{8I_1}{6\frac{I_1}{I}+1} \tag{9}$$

Both formulas, equation (6) determined by Johnston or equation (8) may be used to calculate the critical load for mode B.

When EUROCODE 3 design procedure for compression members is used, these critical loads should be used to calculate the reduction factor χ for the relevant buckling mode of the battened column (mode A or mode B). Rondal and Niazi suggested that the buckling curve for the battened column should be taken according to the component section shape,
so, in this particular case of C sections, the buckling curve **'b'** has to be considered.

The buckling resistance of the battened column, $N_{b,Rd}$, should be computed considering the effective cross-section properties, according to EUROCODE 3 Part 1.3:

$$N_{b,Rd} = \chi A_{eff} f_y \tag{10}$$

in which A_{eff} is the effective area of the entire cross section of the battened member.

EUROCODE 3 method. EUROCODE 3 stipulated that built-up compression members consisting of two or more component sections connected together at intervals to form a single compound member shall be designed incorporating an equivalent geometric imperfection comprising an initial bow of l/500. The deformation of the compound member shall be taken into account in determining the internal forces and moments in the main components.

The chord force $N_{f,Sd}$ at mid-length should be determined from:

$$N_{f,Sd} = 0.5(N_{Sd} + M_s h_0 A_f / I_{eff}) \tag{11}$$

where

$$M_s = N_{Sd} e_0 /(1 - N_{Sd} / N_{cr} - N_{Sd} / S_V) \tag{12}$$

$$e_0 = l / 500 \tag{13}$$

$$N_{cr} = \pi^2 EI_{eff} / l^2 \tag{14}$$

In equation (11) the effective in-plane second moment of area I_{eff} of the battened member should be taken as:

$$I_{eff} = 0.5h_0^2 A_f + 2I_f \qquad \text{if} \quad \lambda = l / i_0 \leq 75 \tag{15}$$

in which A_f is the cross-sectional area of one chord, h_0 is the distance between centroids of chords and i_0 is the radius of gyration for the battened member, computed by means of I_{eff} from equation (15).

Provided that the connections of battens to the chords may be assumed as hinged (as considered in the Rondal and Niazi proposal) the shear stiffness S_V is, in fact, zero. Thus, the chord force $N_{f,Sd}$ from equation (11) became half of the applied force :

$$N_{f,Sd} = 0.5N_{Sd} \tag{16}$$

In case of C cold-formed sections, the ultimate axial load capacity is reached when this chord load, $N_{f,Sd}$, is equal to the buckling resistance $N_{b,Rd}$, computed considering the effective cross-section properties, according to EUROCODE 3 Part 1.3 (1993):

$$N_{b,Rd} = \chi A_{eff} f_y \tag{17}$$

The buckling length of the chord is equal to a. Buckling curve **b** should be considered, in case of C cold-formed sections, as EUROCODE 3 Part 1.3 provides.

Numerical study. A numerical study was performed in order to calibrate a FEM model suitable to predict the ultimate load of built-up columns made of C cold-formed elements connected with bolted C stitches (Dubina et al., 2002). For this purpose, 12 test results from the experimental program performed by Rondal and Niazi (1990, 1993) were considered. Two different C sections with two different lengths of specimens were considered, and three or four C stitches were used to built-up the battened elements. For each set of parameters, three tests were performed. Table 7 shows the specimen characteristics.

The numerical simulations were performed with ANSYS 5.4 program. SHELL43 plastic large strain elements were used for meshing the two C cold-formed chords and the C cold-formed stitches. A bilinear elastic-perfect plastic model for material behaviour was used. Sine shape imperfections of L/1000 have been considered in numerical model, according to ECCS Recommendations (1987), and L/500, according to EUROCODE 3 (1993), respectively. Simultaneously a local geometrical imperfection was introduced Schafer and Pekoz (1998), e.g.:

$$d=0.006\ b \tag{18}$$

where b is the width of relevant wall of C sections.

Table 7. Specimen characteristics

Specimen type	Chords dimensions [mm]	Stitches dimensions [mm]	Specimen length [mm]	Yield limit fy [kN/mm^2]	Number of stitches	Stitches distance [mm]
120.4.4s	C120x60x 18.7x2.4	C80x40x 15x2.5	4000	455	4	1300
120.3.4s	C120x60x 18.7x2.4	C80x40x 15x2.5	3000	455	4	970
180.4.4s	C180x70x 25x2.97	C120x60x 19x2.45	4000	428	4	1300
180.4.3s	C180x70x 25x2.97	C120x60x 19x2.45	4000	428	3	1300

Table 8 shows the comparison between the values of ultimate compression load P_u, computed according to EUROCODE 3 provisions and Rondal and Niazi proposal, respectively, and with those obtained by FEM analysis and tests. For the two design methods, both characteristic and design values P_u/γ_{M1} are given ($\gamma_{M1}=1.1$). For Rondal and Niazi proposal and FEM analysis, the critical loads, $P_{cr,}$ were also computed. The values between brackets in ANSYS column of Table 2 represent the ultimate loads computed with an overall imperfection of L/500, while the other ones are for L/1000.

It may be observed that there is a very good agreement between the critical loads obtained with the Rondal and Niazi proposal and the numerical study. For all specimens, Mode B resulted as critical, both for the design method and numerical study. This is a flexural mode.

Lateral - torsional buckling can be assumed to be restrained both by supporting conditions and stitch fastenings.

Rondal and Niazi proposal gives good results if one compares with experimental values. It may be observed that only a single experimental result, for 180.4.4s specimen, is lesser than the design value, but this value is obviously out of range. Thus, the method can be successfully used in design.

The values obtained with ANSYS program, demonstrate the numerical model is sensitive to the global imperfection. For numerical analysis, a global imperfection of l/500 should be adopted, as recommended by the EUROCODE 3, in order to obtain conservative results.

Table 8. Ultimate and critical loads from design methods, numerical study and tests [kN]

Specimen type	EC 3		Rondal & Niazi proposal			ANSYS		Test		
	P_u	P_u/γ_{M1}	P_{cr}	P_u	P_u/γ_{M1}	P_{cr}	P_u	P_u	m	s
120.4.4s	217	197	251	184	167	264	224 (178)	200 219 188	202	12.8
120.3.4s	385	350	446	268	244	451	341 (232)	345 310 308	321	17
180.4.4s	550	500	599	374	340	598	461 (406)	410 467 330	402	56.2
180.4.3s	437	397	599	374	340	597	460 (405)	438 415 435	429	10.2

EUROCODE 3 approach, dedicated to the battened compression members with plate battens, usually rigidly connected on the chords, leads to securitary results only for specimen 180.4.3s only, which has one intermediate stitch. As demonstrated by Rondal and Niazi method, FEM analysis and experimental results, the number of intermediate stitches does not have a significant influence on the value of ultimate load.

This conclusion is available, of course, in case of flexible stitches. If stiff battens will be used, then, according to the design procedure in EUROCODE 3, the upper limit of shear rigidity, Sv, can be taken, and the design values corresponding to EC3 column in Table 2 would be reduced with 20% around. Therefore, if ticker sections would be used, then the EC 3 procedure can be expected to be applied. Of course, experimental tests and/or numerical simulations are necessary in order to validate this assumption.

8.5 Real behaviour of bolted connections in frames and trusses

Truss joints. Trusses used for WSMS or PFS are of Vierendell type with central braces or trapezoidal type, (see Figure 19 and Figure 20). A typical joint detail is given in Figure 21.

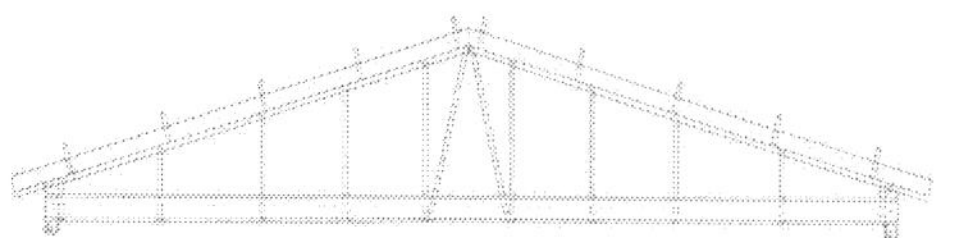

Figure 19. Vierendell truss with central braces

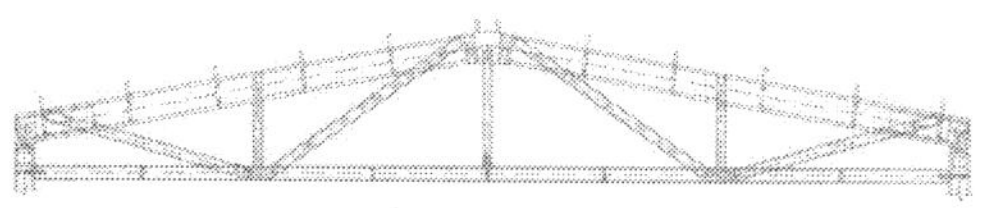

Figure 20. Truss of trapezoidal type

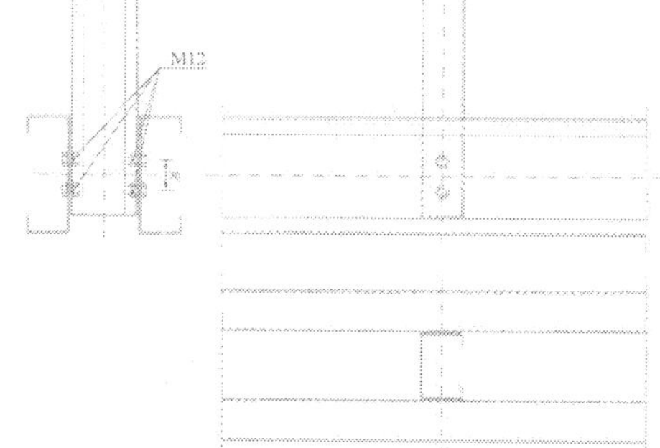

Figure 21. Joint detail

Joints of these trusses are usually with eccentric connections, and this feature must be taken into account in the global analysis. The use of 2, 3 or 4 bolts on each flange of the diagonal or strut members, is supposed to modify the assumption of pinned joints, generally accepted in case of trusses.

Consequently, the real behaviour of joints, in this case, is semi-rigid with partial moment resistance. This has as effect a favourable reduction in the buckling length of diagonals or struts, but at the same time, due to the rigidity and possible eccentricity of connections, supplementary bending moments are induced in members. To evaluate the real characteristics of joints, the flexibility of connection, due to the bearing work of bolt in the thin plate, associated with the hole elongation, bolt tilting and slippage, must be considered. In order to estimate the performance of bolted joints in cold-formed steel trusses, an extensive research programme was developed in the Laboratory of Steel Structures of the "Politehnica" University of Timisoara, and will be further presented.

Portal frame joints. Beam-to-column joints in cold-formed steel portal frames are usually bolted. The joint particular detailing depends of the shape of member sections. For HTMS and PFS buildings, the typical frame joint is shown in Figure 22. For small spans, it may be not necessary to use the stiffening (bracket) bar.

For truss systems the semi-rigidity of joints could represent a benefit, compared with the classical pinned assumption, leading to a reduction of member buckling lengths. Contrary, for portal frames, a rigid joint behaviour is generally assumed in global analysis and the semi-rigidity of the connections may lead to unsafe results. The rotational flexibility of the connections may significantly modify the bending moment distribution in the frame, increases the deformability and second order effects. Consequently, the stiffness of the joints must be evaluated, in order to control the behaviour of portal frames, especially when they are made of cold-formed steel sections.

It must be emphasised that the actual EUROCODE 3 provisions for joint design (pr. EN 1993-1.8, 2001) do not apply the method of components for such type of joints and for time being, design provisions to evaluate their properties are not available.

Tests on knee joints in portal frames were performed by Wilkinson (1999). The aim of the experimental program was to examine the ability of various joint types to form plastic hinges.

Rectangular hollow sections were considered for columns and beams, realising knee joints by means of stiffened or unstiffened welded connections, bolted end plates, bolted or welded internal sleeves.

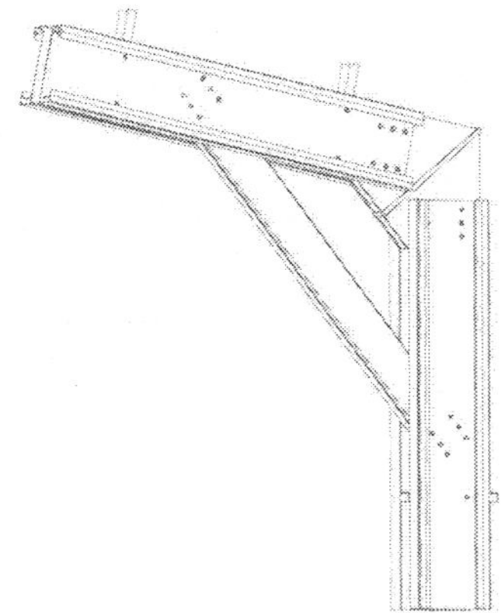

Figure 22. Typical frame joint

Failure modes and moment-rotation curves of these connections were determined. It was decided that the internal sleeve connections exhibited the most suitable behaviour for plastic design. Large plastic rotations, much larger than those of the stiffened connections, were achieved for these joints.

In a recent study by Lim (2001), the behaviour of cold-formed steel portal frames of built-up back-to-back channel - sections (Valleybeam and Swagebeam sections), with bolted moment-connections, was analysed both experimentally and numerically (e.g. FEM simulations). In a first step of this study, three buckling modes of failure of the eaves joints, have been identified, i.e.: overall lateral-torsional buckling of joint, buckling of the stiffened free-edge of the eaves bracket into a one-half since wave and local plate buckling of the exposed triangular area of the eaves bracket. Design recommendations corresponding to the three failure modes are presented.

Experimental study in order to calibrate a formula for the flexibility of single lap bolted connection of two thin plates was systematically performed by Zadanffarokh and Bryan (1992). This formula gives the axial flexibility of a single lap connection in terms of plate thickness and considering the threaded portion of the threaded portion of the bolt into connection. The bolts used in the eaves joints are in double-shear and consequently the bolt - tilting is prevented. Thus, the design formula proposed by Zadanfarrokh and Bryan, including the bolt-tilt is not applicable. A supplementary study was then performed, in order to calculate the rotational stiffness of the joints. A non-linear large displacement elasto-plastic finite element idealisation of a plain bolt-shank in bearing against a bolt-hole was performed and the load-extension characteristic for bolt-hole elongation was determined.

Finally, a simple frame model that takes into account the semi-rigidity of the joints is proposed. The results of the beam idealisation were verified against finite element shell analysis and full-scale laboratory tests. The efficiency of the portal frame is compared to that of an equivalent rigid-jointed frame, taking into account for serviceability requirements practical detailing constraints on the size of the brackets and number of bolts used for the joints. The conclusion of the study is that semi-rigid design must be used for cold-formed steel portal frames as the rotational flexibility of the connections has a large effect on the bending distribution of the frame.

The behaviour of some column-base and sub-frames bolted moment connections using hot-rolled gusset steel plates was analysed recently by Wong and Chung (2002). As in the previous study, only the webs of lipped C-sections are bolted onto gusset plates for ease of build-ability. Four bolts per member were used as minimum configuration.

This type of bolted connections was shown to be effective in transmitting moment between the connected sections, enabling effective moment framing in cold-formed steel structures. Four modes of failure were identified, i.e.: bearing failure in section around bolt-hole, lateral - torsional buckling of gusset plate, flexural failure of gusset plate and flexural failure of connected cold-formed steel section. The moment resistances of the proposed connections configurations were found to range from below 50% to over 85% of the moment capacities of the connected sections.

Experimental and numerical studies by Zaharia and Dubina (2000) focused truss joints of the type shown in Figure 27, but their findings can be easily extended to the frames. Therefore, the relevant results of this research will be summarised on the following.

Experimental study on the behaviour of bolted joints in cold-formed steel trusses. The behaviour of bolted joints in cold-formed steel trusses was investigated through an extensive research programme in the Laboratory of Steel Structures of the 'Politehnica' University of Timisoara.

Ten T joints specimens were tested by Zaharia and Dubina (2000). They are shown in Figure 23, while Figure 24 shows the experimental arrangement.

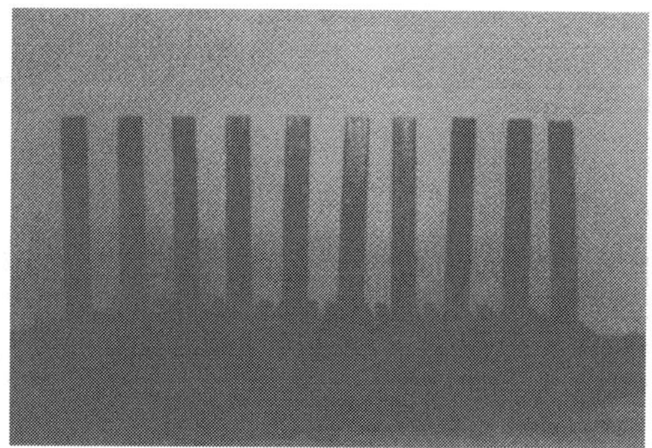

Figure 23. T joint specimens

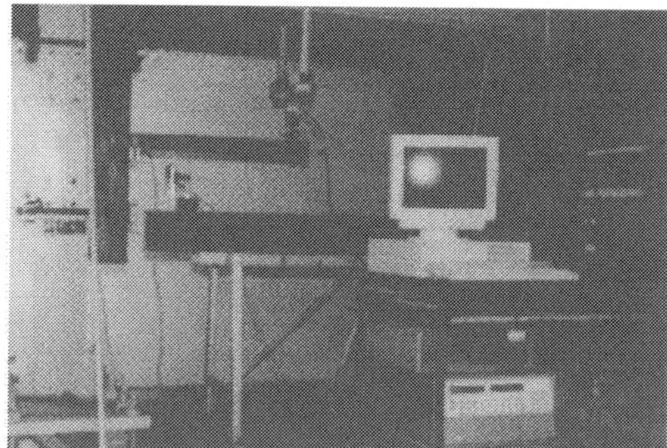

Figure 24. T specimen in testing machine

For all tested specimens, the rotational flexibility was influenced mainly due to the bearing work of bolts. Both elastic and plastic deformation of the bolt hole must be considered. Rotational rigidity of the joint can be evaluated if the axial rigidity of the single lap bolted connection in shear is known. If the initial rotational slippage is neglected, then, according to EUROCODE 3, all tested joints can be classified as semi-rigid and partial moment resisting.

In fact, the triangulated shape of trusses, which is geometrically and kinetically stable, and the presence of the axial forces in connected members, prevent or limit the initial rotational slippage into the joint. In order to provide the evidence of this assumption, a cold-formed steel truss was tested in the next step of this experimental programme (Zaharia and Dubina, 2000).

In order to evaluate the axial rigidity of a single bolt lap joint, in terms of the thickness of plates and bolt diameter, an experimental programme was carried out by Zaharia (2000) who calibrated the following formula:

$$K_{bolt} = 6.8\frac{\sqrt{d}}{\left(\frac{5}{t_1}+\frac{5}{t_2}-1\right)} \tag{19}$$

in which K_{bolt} is the rigidity of a single bolt lap joint, d is the nominal diameter of the bolt and t_1, t_2 are the thickness of joined plates. The ranges of validity of this formula are the bolt diameter between M8-M16 and the thickness of plates between 2-4mm. It can be noticed that the partial safety factor for this formula is identical with the partial safety factor used in EUROCODE 3 Part1.3 for the resistance of bolted connections, e.g. $\gamma_{M2} = 1.25$.

Computation models for the rotational rigidity of T-joints were established, depending of the axial rigidity of single bolt lap joint. The theoretical model for two bolts joint proves a good correlation of results obtained from tests.

The computation scheme for the rotational rigidity of a diagonal-to-chord joint, with two bolts on each flange of the diagonal, is shown in Figure 25.

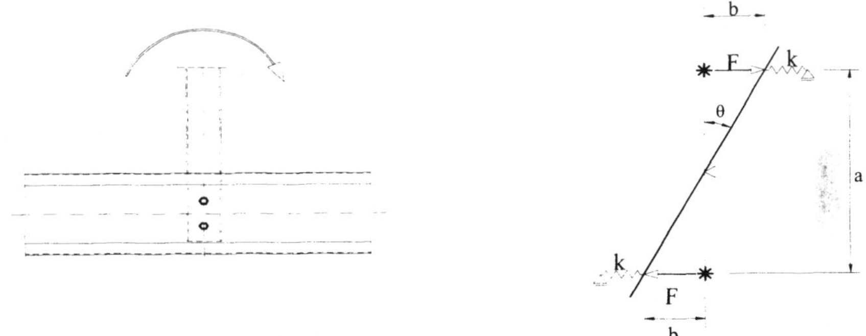

Figure 25. Computation model for two bolts joint

Using the notations from Figure 25, the rotational rigidity of the joint, K_{node} can be expressed in terms of total bending moment and corresponding rotation, θ, as (Zaharia, 2000):

$$K_{node} = \frac{M_{tot}}{\theta} = \frac{2kda}{\left(\frac{d}{0.5a}\right)} = ka^2 = \frac{6.8a^2\sqrt{d}}{\left(\frac{5}{t_1}+\frac{5}{t_2}-1\right)} \quad \text{[kNmm/rad]} \tag{20}$$

in which

$$F = k \times b \tag{21}$$

$$tg\theta = \theta = \frac{b}{0.5a} \tag{22}$$

In order to prove the initial rotational slippage observed in case of tested joints is not significant, when the joint is working in the truss structure, and to validate the theoretical assumptions introduced above, a full-scale test of a truss specimen was performed (Zaharia and Dubina, 2000). The dimensions of the experimental model are shown in Figure 26. The experimental arrangement is shown in Figure 27. All connections are made with M12, 8.8 grade bolts.

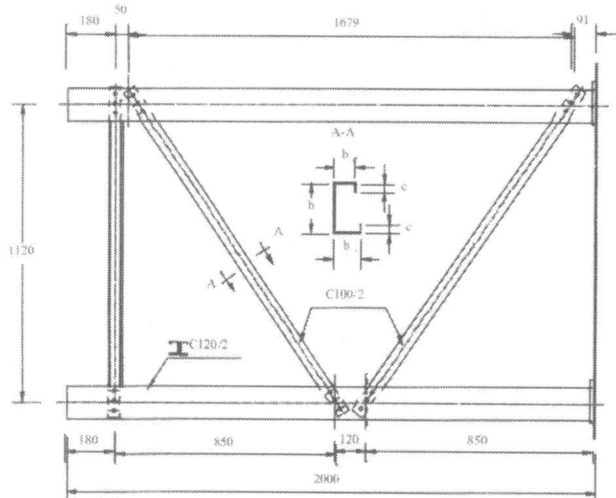

Figure 26. Experimental model

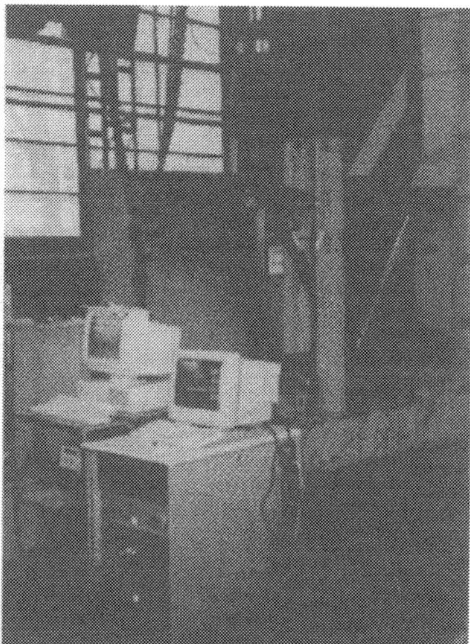

Figure 27. Experimental arrangement

The load increased until the structure failed due to the flexural instability of the diagonal in compression, in the plane of the truss. A local buckling in the lower chord, due to the shear of the web panel located between diagonals, was also observed before the buckling of diagonal was reached. This phenomenon contributes to the deformability of the joint, too.

Figure 28 presents the axial displacements reported by the LVDT transducers. It can be observed the typical behaviour of a thin plate bolted connection in shear. After the attainment of load corresponding to the initiation of slippage in connection, this can be developed until the hole clearance is consumed. Figure 29 shows the evolution of the diagonal rotations. Corresponding to the load range in which the axial slippage occurs, very small rotations are observed only. Until the structure 'shake down', the presence of the axial forces and the triangulation effect prevent the developing of significant rotational slippage in connections. Practically, the initial rotational slippage, observed during the test of joints, is almost totally restraint into structure. Consequently, the rotational rigidity evaluated without considering the initial slippage, is a real one and can be used in the global analysis, and to evaluate the buckling length of relevant members.

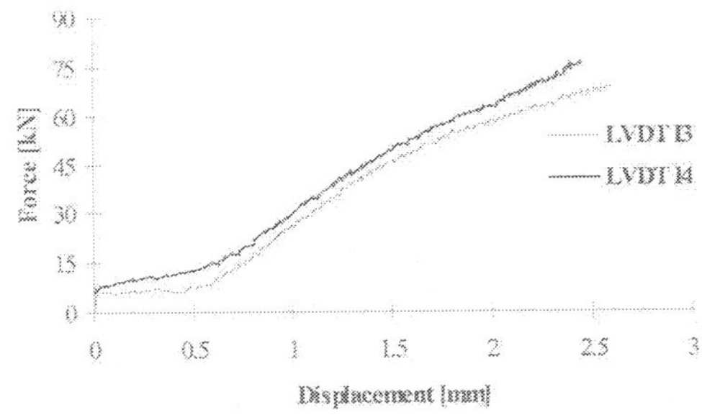

Figure 28. Axial displacements of the diagonals

Figure 29. Diagonal rotations

Concluding remarks. As stated beginning of this sections, in most cases in the behaviour of joints in cold-formed trusses and frames are semi-rigid. This fact was proved both experimentally and numerically. The designer must take care to control this phenomenon using proper design procedures, as design assisted by tests.

8.6 Seismic performance of light-gauge steel structures

8.61 Generals

According to EN 1998, EUROCODE 8 (2001) a seismic resistant structure shall be designed in such a way, that following fundamental requirements are meet, each with an adequate degree of reliability:

- No collapse requirement;

The structure shall be designed and constructed to withstand the seismic action defined in Section 3 without local or global collapse, thus retaining its structural integrity and a residual load bearing capacity after the seismic events. The reference seismic action is associated with a reference probability of exceedance in 50 years and a reference return period.

- Damage limitation requirement.

The structure shall be designed and constructed to withstand a seismic action having a larger probability of occurrence than the seismic action used for the verification of *the "no collapse requirement"*, without the occurrence of damage and the associated limitations of use, the costs of which would be disproportionately high in comparison with the costs of the structure itself. The reference seismic action to be taken into account for the 'damage limitation requirement' has a low probability of exceedance in 10 years. In the absence of more precise information, the reduction factor q (see paragraph 4.4.3.2. in the code) may be used to obtain the seismic action for the verification of the 'damage limitation requirement'.

Target reliabilities for the *"no collapse requirement"* and for the *"damage limitation requirement"* are established by code provisions for different types of buildings or civil engineering works on the basis of the consequence of failure.

Reliability differentiation is implemented by classifying structures into different importance categories. To each importance category an importance factor γ_I is assigned. Wherever feasible this factor should be derived so as to correspond to a higher or lower value of the return period of the seismic event (with regard to the reference return period), as appropriate for the design of the specific category of structures.

The different levels of reliability are obtained by multiplying the reference seismic action or - when using linear analysis - the corresponding action effects with this importance factor. Detailed guidance on the importance categories and the corresponding importance factor is given in the relevant Parts of EUROCODE 8 (1998).

In order to satisfy the fundamental requirements set forth above, the following limit states shall be checked:

- *Ultimate limit states,* are those associated with collapse or with other forms of structural failure which may endanger the safety of people.
- *Serviceability limit states,* are those associated with damage occurrence, corresponding to states beyond which specified service requirements are no longer met.

A seismic resistant structure should be provided with balanced stiffness, strength and ductility between its members, connections and supports (Bertero, 1997). Conceptually, Seismic Resistant Steel (SRS) structures are always redundant, because redundancy is the inherent condition of the reliability of structural systems. Therefore, in the American Design Code (AISC, 1997), redundancy (or reliability) coefficient is a component of reduction factor R (corresponding to behaviour q factor in EUROCODE 8). Bertero R. and Bertero V. (1999) have shown the redundancy (as well as reduction or behaviour factor) is dependent of overstrength and ductility of structural system.

To comply with ULS criteria, the structure shall be verified in terms of resistance and ductility. The resistance and ductility to be assigned to the structure are related to the extent to which its non-linear response is to be exploited. In operational terms such balance between resistance and ductility is characterised by the values of the behaviour factor q, which are given in the relevant parts of EUROCODE 8. As a limiting case, for the design of structures classified as non-dissipative, no account is taken of any hysteretic energy dissipation and the behaviour factor is equal to 1.5, provided the structure has some overstrength.. For dissipative structures, the behaviour factor is taken greater than 1.5 accounting for the hysteretic energy dissipation that occurs in specifically designed zones, called *"dissipative"* or *"critica"l* regions.

To comply with SLS criteria, expressed in terms of deformation limits, the structure shall be verified to possess sufficient resistance and stiffness. There are some general design criteria which apply for all structures located in moderate and severe seismic zones, i.e.:

a) Structures should have simple and regular forms both in plan and elevation. If necessary this may be realised by subdividing the structure by joints into dynamically independent units.

b) In order to ensure an overall ductile behaviour, brittle failure or the premature formation of unstable mechanisms shall be avoided. To this end, it may be necessary to resort to the capacity design procedure, which is used to obtain the hierarchy of resistance of the various structural components and failure modes necessary for ensuring a suitable plastic mechanism and for avoiding brittle failure modes.

c) Since the seismic performance of a structure is largely dependent on the behaviour of its critical regions or elements, the detailing of the structure in general, and of these regions or elements in particular, shall be such as to maintain under cyclic conditions the capacity to transmit the necessary forces and to dissipate energy. To this end, the detailing of connections between structural elements and of regions where non-linear behaviour is foreseeable should receive special care in design.

d) In order to limit the consequences of the seismic event, National Authorities may specify restrictions on the height or other characteristics of a structure depending on local seismicity, importance category, ground conditions, city planning and environmental planning.

e) The analysis shall be based on an adequate structural model, which, when necessary, shall take into account the influence of soil deformability and of non-structural elements and other aspects, such as the presence of adjacent structures.

f) Special care must be payed for correct design of foundations accounting for the local conditions of the soil and structure typology.

Cold-formed steel structures are usually made by thin-walled sections, of class 4 or, at most, of class 3. Compared with hot - rolled sections (of class 1 or 2), they are characterised by a reduced post-elastic strength and, as a consequence, by a reduced ductility (e.g. they do not have sufficient plastic rotation capacity to form plastic hinges, as shown in Figure 29).

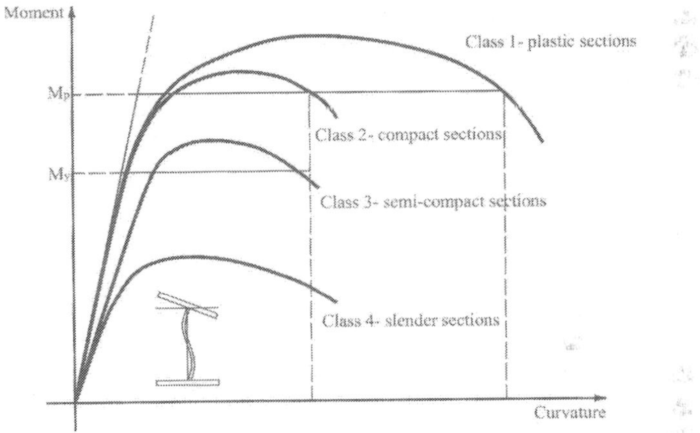

Figure 29. Cross-section behaviour classes (Gioncu and Mazzolani, 2002)

According to EN 1998 provisions, a cold-formed steel structure must be classified as a non-dissipative structure, thus the behaviour factor q will be taken 1.5 for current design. In fact, the seismic design of cold-formed steel structures must be based on strength criteria and overall rigidity control, because no ductility criteria may apply.

The sources of robustness of these structures under seismic actions are mainly based on redundancy and overstrength. Consequently, a proper conceptual design is crucial for their performances.

The European specific design rules for cold-formed steel design (e.g. EN 1993-1-3, 2001) have no recommendations for seismic design of these structures. In the new North American Specification (AISI, 2001), Section *G, "Design of cold-formed steel structural members and connections for cyclic loadings"*, rather reefers to fatigue then to seismic behaviour. Thus, actually cold-formed steel design codes do not contain specific recommendations for seismic design of cold-formed steel structures.

8.6.2 Behaviour factors. Ductility and overstrength.

To explain the physical background of behaviour factor, also called reduction factor in seismic design codes, and to understand the ductility and overstrength contributions to the structural performance, the Chapter 8.1 *"General definitions and basic relations"* by P. Fajfar in the book *"Moment Resistant Connections of Steel Frames in Seismic Areas. Design and Reliability"* edited by F.M. Mazzolani (2000) will be taken as basic reference and summarised hereafter.

Considering the structural system simulated by an equivalent single degree-of-freedom (SDOF) model with an idealised bilinear force-displacement behaviour (Figure 30) the following relations apply:

$$F_y = m\,\omega^2 D_y \tag{23}$$

$$\mu = \frac{D}{D_y} \tag{24}$$

where F_y is the strength of the system (yield strength) , m its mass, ω the natural frequency, D_y the yield displacement, D the maximum displacement, and μ the corresponding ductility factor.

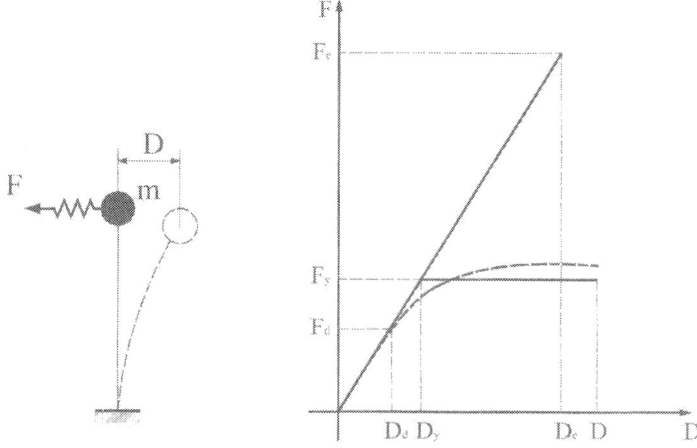

Figure 30. Idealised force-displacement relationships for a SDOF system

A structure with a given strength F_y, which relies on energy dissipation trough inelastic deformation, should have a limit deformation capacity, which exceeds seismic demand in the case of severe earthquakes. It is well known that seismic demand in terms of ductility (which is related to deformation according to Eqn. 24), and seismic demand expressed in terms of elastic strength F_y are interrelated. The problem can therefore be stated in a different way. Assuming that a certain deformation capacity is provided (or, in the case of serviceability limit state, a certain deformation is accepted), the strength F_y of the system should, at least, be equal to the required strength. This approach is actually used in design procedures (*force-controlled design*) and can be written in the form:

$$F_y = \frac{mA_e}{q_\mu} \tag{25}$$

where A_e is the amplitude value in the elastic (pseudo) acceleration spectrum and q_μ is the strength reduction factor, which is equal to the elastic strength demand $F_e = m\ A_e$ divided by the inelastic strength demand F_y:

$$q_\mu = \frac{F_e}{F_y} \tag{26}$$

q_μ depends mainly, but not exclusively, on the prescribed target ductility and on the period of the system. Expressions similar to equation 25 can be found in various seismic codes. However, an important difference should be noted between equation 25 and the expressions in the codes. In Eqn. 25, F_y represents the actual strength, whereas the seismic forces in codes correspond to design strength F_d which is, as a rule, lower than the actual strength. This difference is mainly due to overstrength, which is an inherent property of well-designed, detailed, constructed and maintained redundant structures.

Taking into account the overstrength factor is

$$q_s = \frac{F_y}{F_d} \tag{27}$$

the following applies

$$q = \frac{F_e}{F_d} = \frac{F_e}{F_y}\frac{F_y}{F_d} = q_\mu q_s \tag{28}$$

Thus, the total force reduction factor q, which is equal to the elastic strength demand F_e divided by the code prescribed seismic force F_d, can be defined as the product of the ductility dependent factor q_μ and the overstrength factor q_s.

The ductility factor q_μ depends on the structure type and the local ductility of the cross-sections and members, level and type of loadings etc. The overstrength factor may come from the internal force distribution, higher material strength than those specified in the design, strain hardening, member oversize, strain rate, multiple loading combinations, effect of non-structural elements, etc. Having in mind that the ductility factor decreases in the short period region, while the overstrength factor increases in this region, from the point of view of the structural design practice, it appears to be a reasonable approximation to use a constant period independent behaviour factor q. (Figure 31).

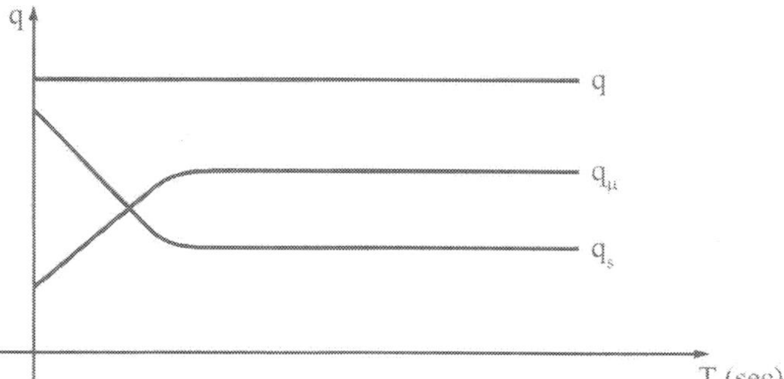

Figure 31. Reduction factors q, q_μ and q_s

The code suggested values of q factors are essentially of an empirical origin. So, in addition to ductility, they generally automatically imply overstrength, although this is usually not explicitly realised. According to EN 1998 (2001), the behaviour factor q is an approximation of the ratio of the seismic forces, that the structure would experience if its response was completely elastic with 5% viscous damping, to the minimum seismic forces that may be used in design - with a conventional linear model - still ensuring a satisfactory response of the structure. Figure 32 shows with reference to SDOF elasto-plastic system of Figure 1 the relationships between elastic, actual and design spectra (Gioncu and Mazzolani, 2002).

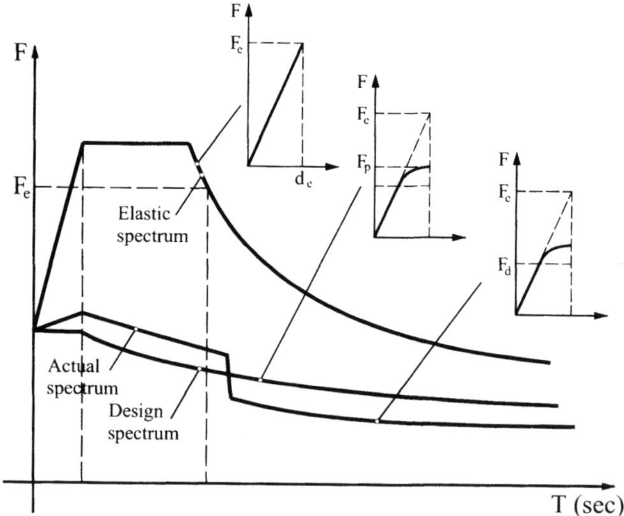

Figure 32. Elastic, actual and design spectra

Usually, the majority of codes use constant behaviour factors throughout the entire spectrum. In EN 1998, q factor is expressed of α_u / α_1 ratio, where:

α_1 - is the multiplier of the horizontal seismic design action which, while keeping constant all other design actions, corresponds to the point where the most strained cross-section reaches its plastic resistance. In case of hot-rolled sections of class 1, it corresponds to the occurrence of first plastic hinge. For cold-formed steel sections, which are generally of class 4, this fact is not possibly; in such a case, the occurrence of local-plastic mechanism may be assumed as reference for ULS plastic resistance of relevant member.

α_u - is the multiplier of horizontal seismic design action, which, while keeping constant all design actions, corresponds to the point where a number of sections, sufficient for the development of overall structural instability, reach their plastic moment resistance.This factor may be obtained from a geometric first order global inelastic analysis (rigid plastic).

Again, for the ULS criterion, in case of cold-formed virtual sections, the local plastic mechanisms may replace the plastic hinges. The drift limitation can be also used to define the *"no collapse requirement"*, in order to avoid the elastic - plastic instability or *"damage limitation requirement"*.

In case of a given structural typology, draft limitation values for ULS can be obtained using a performance based design philosophy (Ghobarah, 2001), and the displacement controlled analysis. Probably the best way is to use the pushover analysis.

According to ENV 1998 (2001), the displacement demand can be computed as

$$D_s = q_d D_e \tag{29}$$

where

D_s - is the displacement of a relevant point of the structural system induced by the design seismic action;

q_d - is the displacement behaviour factor, assumed equal to q values otherwise specified;

D_e - is the displacement of the same point of structural system, as determined by a linear analysis based on the design response spectrum.

If an equivalent static elastic-plastic (e.g. pushover) analysis is performed, q factor can be also evaluated with the following formula (Mazzolani and Piluso, 1996):

$$q = \frac{\alpha_u}{\alpha_1}\left[\left(1-\beta'\right)\alpha_{cr} + \beta'\right] \tag{30}$$

where

$\beta' = 1 - T$; $\beta' \geq 0.5$

T - is the fundamental period of structure;

α_{cr} - is the critical load multiplier of gravitational loads, V (e.g. $\alpha_{cr} = V_{cr} / V$);

For SLS, assuming the cold-formed steel-framed building has ductile cladding elements, the EN 1998 drift limit value is:

$$D_r\, v \leq 0.0075\, h \tag{31}$$

where

D_r - design interstorey drift;

h - is the storey height

v - is the reduction factor to take into account the lower return period of the seismic event associated with SLS.

This value is really to severe for the actual materials and for cladding in light gauge steel buildings (see also Section 8.6.7).

8.6.3 Overstrength

As already stated, the overstrength is source of redundancy in case of non-dissipative structures, as cold-formed steel one are, even some ductility supply is inherent. Thus, when such structures are designed, the so called Strength Design Criterion (SDC) has to be applied. SDS compares the Required Strength (RS) with the Available Strength (AS), e.g.

$$\mathrm{RS} \leq \mathrm{AS} \tag{32}$$

RS depends of structure typology, earthquake type and design spectrum. The difference between AS and RS, representing the strength exceeding that required by code (RS is the overstrength).

According to Paulay and Priestley (1992), the overstrength factor is basically defined at the section (of a member) level as the ratio between the maximum strength and the nominal or theoretical strength. The nominal strength of a section is based on the established theory predicting a prescribed limit state with respect to failure of this section. It is derived from the dimensions and code-specified nominal material strength properties. Factors that may contribute to strength exceeding the nominal value include actual steel strength greater than the specified yield strength, strain hardening and strain rate effects. The overstrength factor according to this definition is typically relatively small and its quantification is difficult or impossible.

Other overstrength factors are also used by the same authors, e.g. flexural overstrength factor and system overstrength factor. The flexural overstrength factor measures the flexural overstrength in terms of the required strength for earthquake forces alone at one node point of the structural model. The system overstrength factor represents the sum of the overstrengts of a number of interrelated members (all definitions are taken from Paulay and Priestley,1992).

According to several other authors (Fajfar, 2001), the overstrength factor is defined on the level of the whole structure as a ratio between the actual structural yield level (note that, in actual structures, modelled as a MDOF system, this is not the first yield level, but the yield level in an idealised bilinear force - deformation diagram) and the code prescribed strength demand arising from the application of prescribed loads and forces. It results from the following groups of sources:

a) redistribution of internal forces in the inelastic range in the case of ductile, statically indeterminate (redundant) structures; difference between the design level and required member strength (e.g. allowable versus yield stresses, load factors); member oversize (due to discrete member sizes and due to desired uniformity of members for constructibility); minimum requirements according to code provisions regarding proportioning and performance; architectural considerations;

b) conservatism in mathematical models; effects of structural elements that are not considered as a part of lateral load resisting system; effects of non-structural elements

c) higher material strength than those specified in the design, strain hardening and strain rate effects.

The influence of the majority of group (a) factors can be easily, at least, approximately quantified by a non-linear push-over analysis, while the (b) group sources are less reliable or require sophisticated mathematical modelling and may be neglected in practical design. The c) group factors are the same as those determining the basic overstrength factor as defined by Paulay and others, and they are difficult to be quantified.

It is obvious that overstrength may come from a variety of sources and that, in real structures, it varies widely, depending on the material and the type of the structural system, the structural configuration, the number of stories, detailing, and the kind and the date of the code to which the structure was designed (Fajfar, 2000).

8.6.4 Ductility of thin-walled sections

Thin-walled sections are prone to local buckling and for this reason they are considered non-ductile (see Figure 1). However, considering the interaction local buckling modes leads to localisation of buckling pattern, the failure of a thin-walled steel section always occurs by a local plastic mechanism. Therefore, even it is small, some amount of post-elastic strength for such a type of section can be considered.

Dubina and Ungureanu (2000) used the local plastic mechanism theory developed by Murray and Khoo (1981) to characterise the local and interactive buckling of cold-formed steel sections. Moldovan et al. (1999) used the same theory to evaluate the plastic rotation capacity (e.g. ductility) of built-up U and C sold-formed steel beams. They developed the DUCTROT - TWM computer code, calibrated via tests results obtained at the University of Naples (De Martino et al, 1992) and found an available cross-sectional ductility, μ, equal to 1.7

$$\mu = \frac{\chi_u}{\chi_y} \tag{33}$$

where χ_y is the beam curvature corresponding to the initiation of plastic deformations, while χ_u is the ultimate curvature.

Based on these results, the authors suggested to use a behaviour factor q=1.7 when designing steel structures made by such type of sections.

8.6.5 Seismic performance of light gauge portal frames

Extended research on this subject was performed by Calderoni and co-authors (1995; 1997). The authors started their research from the experimental findings of Ono and Suzuki (1986) who proved by tests significant post-elastic strength and ductility of some cold-formed steel frames. Calderoni et al. (1997) characterised the monotonic behaviour of portal frames by assuming a simplified F-D curve, constituted by a bi-linear increasing branch, a linear softening branch and an horizontal constant one for the residual bearing capacity, while the cyclic behaviour was fixed by means of some degradation rules, for both stiffness and strength in post-elastic range plastic excursion (Figure 33). By using this kind of cyclic load-displacement law, a lot of numerical step by step dynamic analyses were performed with reference to some built-up channel section portal frames. Geometrical and mechanical properties of frames were selected to provide monotonic F-D curves characterised by elastic stiffness, slope of the softening branch, and residual strength.

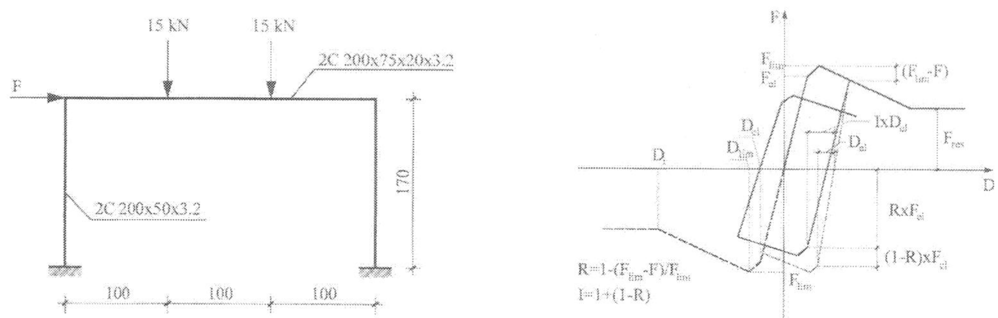

Figure 33. Analysed frame and analytical cyclic behaviour

The dynamic response of the analysed frames (performed by taking into account also the geometrical degradation due to second order effects) was obtained by using, as load conditions, thirty different real accelerograms, recorded during some Italian earthquakes. They were selected in such way that the corresponding average elastic response spectrum (50% of probability to be exceeded) to be similar to that provided by EN 1998 for soil type A and PGA equal to 0.15g.

The results of this wide numerical investigation (about 1000 analyses) showed that the seismic behaviour of thin-walled portal frames was not so different on respect of the corresponding ideal elasto-plastic structure, provided that the slope of softening branch of the monotonic F-D curve is reasonable (e.g. 30° around). In these cases, it seemed that a q factor greater than 1, varying in the range 2 to 3, could be used in low-seismicity zones, if the available ductility exhibited by the frame is about equal to 3 (Calderoni et al., 1994). Nevertheless it clearly appeared that the shape of the F-D curve, the lateral elastic stiffness, affect the structural response and consequently the judgement on the possibility of using the light gauge structures in seismic zones. However, the lowest limit of q factor, suggested by the authors of this study is 1.8 (the q value equal to 3, for the frame of Figure 5, is really to big to be used in practice, since the sections are unusually thick). This value approaches to the previous one proposed by Moldovan et al. (1999), and both are practically of the same order of magnitude with the value given in EN 1998 (2001) for non-dissipative structures, e.g. q=1.5.

It is important to emphasise that a significant improvement of seismic performances of cold-formed steel framing can be obtained if the co-operation with structural envelopes is considered. This matter will be analysed in Section 8.6.7.

8.6.6 Strength and ductility of shear walls

Generals. Review of existing research results. Wall stud based shear walls can be used both on residential or non-residential applications of cold-formed steel sections as well as for industrial buildings. Their behavior and design under static loading can be managed on the base of ECCS Recommendations (1995). Their strength and ductility properties are important for buildings which have to carry significant horizontal loading from wind and earthquake action. Even if widely used in practice, the behaviour of shear walls subjected to earthquake is not fully clarified by research. In recent years important progress has been made to understand certain aspects related to shear strength, stiffness and ductility.

Research in the US (AISI 1997, Serette 1996, Salenicovich 2000) has been focused mainly towards experimental testing of shear walls typical to their home practice, in order to produce practical racking load values. Load bearing capacities are derived both from monotonic push-over curves, envelope and stabilised envelope curves from cyclic tests. Findings of these studies suggest a conventional elastic stiffness for a wall panel at 0.4 of the ultimate load (Serette 1998). Different frame typologies with various cladding materials were tested, studies being conducted to determine the influence of length/height ratios as well as the effect of openings. Even if very detailed, the majority of studies avoid addressing an important aspect of shear wall behaviour, energy dissipation capacity due to cyclic characteristics (Serette, 1996). The effect of gypsum wallboard was also studied, leading to the conclusion that both strength and stiffness are increased by the presence of gypsum wallboard, some results suggesting an increase in ultimate load of up to 30%, compared to the case of external sheeting.

Testing and numerical simulation was combined in order to account for hysteretic characteristics in an attempt to provide evidence on the possible values of response modification factors 'q' (Kaway 1999). Vibration tests of steel-framed houses were conducted and relatively large damping ratios were found due to interior and exterior finishes. According to the tests damping ratio of 6% was accepted for seismic analysis. A maximum 1/50 rad story drift angle limit is also suggested as acceptable during severe earthquakes. In FE analysis stage, a steel-framed house was subjected to two levels of seismic waves. The house exhibited good performance, reaching a maximum drift of 1/300 rad. Even when minimum required wall length was provided, the maximum drift did not exceed 1/60 rad.

The same issue is analysed by Gad (Gad et al. 2000, Gad 1999), who proposes a new analytical approach to evaluate the ductility parameter (R_μ), and finds a value between 1.5 and 3.0 to be suitable. The same research briefly assesses inherent structural overstrength and finds it to be very important factor as far as earthquake resistance is concerned. The quantitative evaluation of overstrength is more difficult, but an empirical evaluation attempt is performed in the report. (Gad et al. 2000)

Experimental tests and FE modelling was employed by G. de Matteis (1998) to assess shear behaviour of sandwich panels both in single story and multi-story buildings. A number of six monotonic and six cyclic tests were performed on full-scale sandwich panel specimens of different configurations. In the final stage of the study, dynamic modelling on panels integrated in building structures, under real earthquake records is performed. According to the conclusions diaphragm action can replace classical bracing solutions only in low-rise buildings, and in areas of low seismicity. For multi-story frames cladding panels can only be used in an integrated system, sharing horizontal force with frame effect.

An important aspect of experimentation was to define acceptable damage levels and relate it to the performance objectives. Recent performance objective proposals are based on three or four generally stated goals (Rosowsky 2002): (1) serviceability under ordinary occupancy conditions: (2) immediate occupancy following moderate earthquakes; (3) life safety under design-basis events; (4)collapse prevention under maximum considered event.

Description of the testing program at 'Politehnica' University of Timisoara. The experimental program (Fulop and Dubina, 2002) was based on six series of full-scale wall tests with different cladding arrangements based on common practical solutions in both housing and SIB (Table 9). Each series consisted of identical wall panels, tested statically both monotonic and cyclic. The main frame of the wall panels were made of cold-formed steel elements, top and bottom tracks were U154/1.5, while studs were C150/1.5 profiles, fixed at each end to tracks with two pair SPEDEC SL4-F-4.8x16 (d=4.8 mm) self-drilling self-taping screws. In specimens using corrugated sheet as cladding the sheets were placed in horizontal position, with a useful width of 1035mm and one corrugation overlapping and tightened with seam fasteners SL2-T-A14- 4.8x20 (d=4.8 mm) at 200 mm intervals (Table 9). Corrugated sheet was fixed to the wall frame using SD3-T15-4.8-22 (d=4.8 mm) self-tapping screws, sheet ends being fixed in every corrugation, while on intermediate studs at every second corrugation. Additionally on the 'interior' side of specimens in Series II, 12.5mm thick gypsum panels (1200x2440mm) were placed vertically and fixed at 250mm intervals on each vertical stud.

Table 9. Description of wall specimens

Ser.		Exterior cladding	Interior cladding	Testing method	Load vel. (cm/min)	No. test
O		-	-	Monotonic	1	1
I		Corr. Sheet LTP20/0.5	-	Monotonic	1	1
				Cyclic	6;3	2
II		Corr. Sheet LTP20/0.5	Gypsum Board	Monotonic	1	1
				Cyclic	6;3	2
III		-	-	Monotonic	1	1
				Cyclic	3	1
IV		Corr. Sheet LTP20/0.5	-	Monotonic	1	1
				Cyclic	6 ;3	2
OSB I		10 mm OSB	-	Monotonic	1	1
				Cyclic	3	1
OSB II		10 mm OSB	-	Monotonic	1	1
				Cyclic	3	1
		Total Number of Specimens				15

Bracing was used in three specimens, by means of 110x1.5 mm straps on both sides of the frame. Steel straps were fixed to the wall structure using SL4-F-4.8x16 and SD6-T16-6.3x25 self-drilling screws, the number of screws being determined to avoid failure at strap end fixings and facilitate yielding. 10 mm OSB panels (1200x2440mm) were placed in similar way as the gypsum panels in earlier specimens, only on the 'external' side of the panel and fixed to the frame using bugle head self-drilling screws of d=4.2mm at 10.5 cm.

The main outputs of the experiments were shear force versus horizontal displacement at the top of the wall-specimens. Furthermore, horizontal slip at the base of the wall and uplift displacement was measured in the two corners. As in case of the panels clad with corrugated sheet, the seams govern the failure, relative slip between two steel sheets was also recorded. Load versus lateral displacement curves are presented for all tested specimens (Figure 34), and in order to illustrate monotonic to cyclic results, stabilized envelope curves are being also presented for the cyclic curves.

As seen wall-panels exhibited very complex, and highly non-linear behaviour. In order to evaluate specific properties like elastic modulus, ultimate force or ductility, curves have been interpreted according to an established procedure. The method has been reported by Kawai (Kawai et al. 1997).

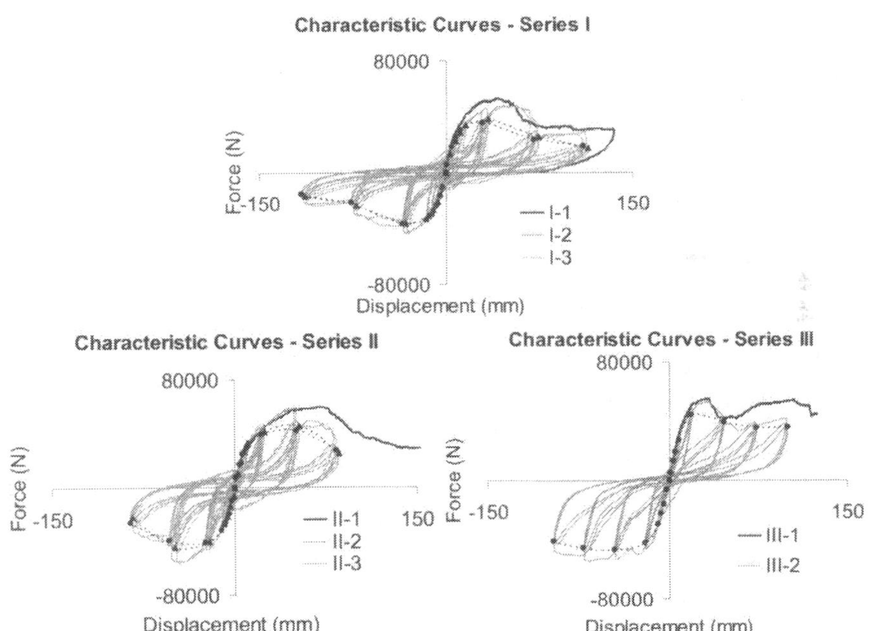

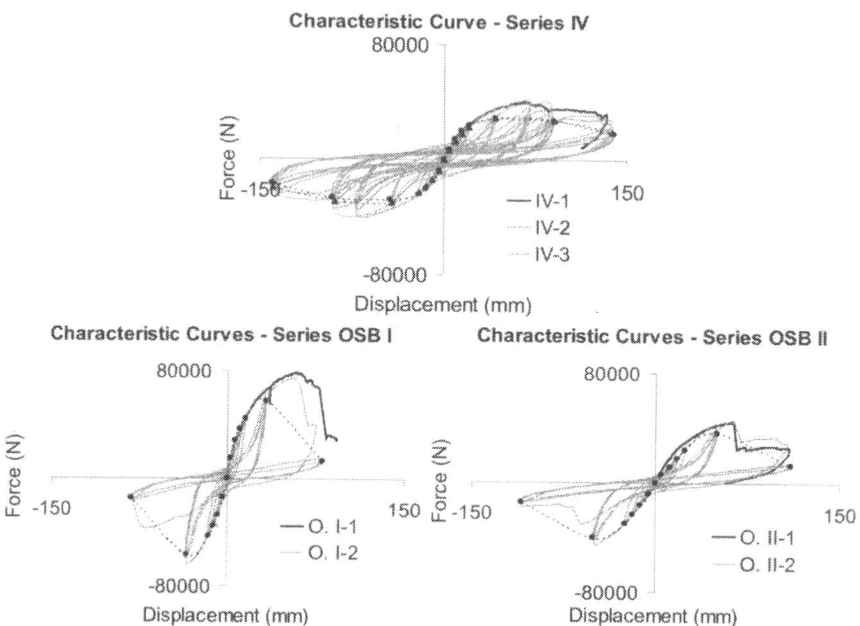

Figure 34. Experimental curves for all specimens

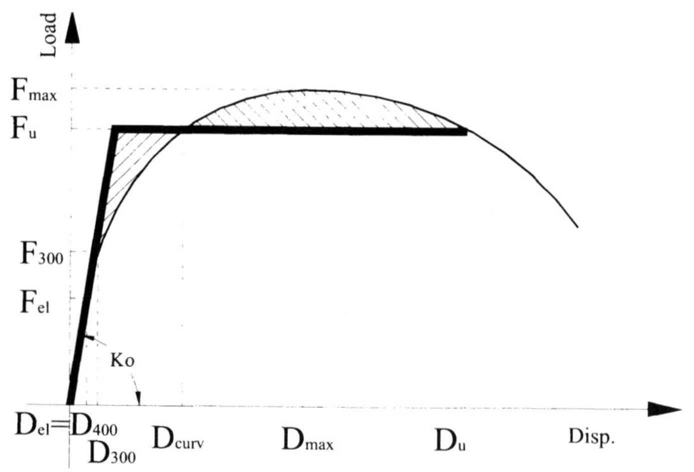

Figure 35. Method for determining equivalent elasto-plastic model

Initial stiffness is defied as the secant stiffness to the point of drift angle corresponding to 1/400 (D_{400}), while the yield line is chosen in a way that the hatched parts in Figure 35 have the same area. The allowable strength is referred as the minimum of the force at story drift angle 1/300 (F_{300}) and 2/3 F_{max}.

Results from monotonic and cyclic experiments are given in Table 10; where two cyclic tests were performed the values are based on the mean value of the two. For cyclic tests, values are derived based on 1st envelope curve (envelope curve), and the 3rd envelope curve (stabilized envelope).

For design capacity, the minimum of $2/3F_{max}$ and F_{300} and F_{200} are relevant, since they represent the values accepted in the Japanese and US code respectively (Kawai 1997).

Differences between monotonic and cyclic values can be observed as follows. Initial rigidity is not affected, values of cyclic and monotonic tests range within a difference of less than 20%. The same can be noted for ductility, exception being in case of OSB specimens where ductility is reduced by 10-25% for cyclic results. One important observation concerns ultimate load (F_u), where cyclic results are lower than monotonic ones by 5-10% even if we consider 1st envelope curve. If we take into account stabilized envelope curves, the difference can increase to 20-30%.

Table 10. Summary of experimental results

Series	Curve	K_o (N/mm)	F_{400}(N)	F_{300}(N)	D_{curv} (mm)	F_u (N)	D_{uct}	2/3 F_{max}	F_{200} (N)
I	Mon	4088.1	24467.2	28690.95	24.8	47821.0	4.65	35250.91	35110.87
	1st Cycle	3446.7	20536.2	24086.55	22.5	39696.6	5.13	30430.85	29425.42
	3rd Cycle				14.9	33560.3	4.37		
II	Mon	3311.5	20088.5	24349.89	41.0	53801.3	5.03	39810.27	30508.56
	1st Cycle	3850.7	22904.4	26565.66	29.5	49029.8	5.05	38286.74	34024.80
	3rd Cycle				15.6	39818.6	5.54		
III	Mon	4187.5	25120.4	31980.04	19.2	51139.6	2.81	36765.07	40192.96
	1st Cycle	3626.8	21259.1	26956.51	21.5	48013.7	5.25	35682.62	36097.14
	3rd Cycle				14.8	42252.2	5.02		
IV	Mon	1598.3	9349.6	13723.80	41.7	35532.7	3.79	26813.45	18048.44
	1st Cycle	1766.3	10416.4	12869.81	37.2	33267.2	5.62	25163.27	17277.07
	3rd Cycle				23.8	26812.2	6.22		
OSB I	Mon	3909.6	23797.3	28470.15	37.2	68162.0	4.26	52517.67	37953.88
	1st Cycle	4197.3	24644.6	28942.46	21.4	54615.2	3.88	46562.66	37194.10
	3rd Cycle				17.5	48944.5	3.67		
OSB II	Mon	1814.9	10702.5	13779.63	36.4	37014.8	3.19	29586.53	18732.50
	1st Cycle	1610.5	9511.3	11850.46	32.3	37426.0	2.93	30539.23	16495.42
	3rd Cycle				27.8	33908.3	3.11		

Based on the medium value of monotonic and cyclic results comparison of different series contribution of opening, gypsum board and other factors can be assessed, with the following conclusions:

Series I - Series II: Differences can be attributed to the effect of the gypsum board. There is an increase of in ultimate load of 16.2% and 17.8% respectively. As far as initial values are concerned (K_o, F_{400}, F_{300}, F_{200}) there seem to be no differences, but ductility is also improved slightly.

Series I – Series IV: There is significant decrease of initial rigidity (60.3% – 53.3%), for a lesser degree of ultimate load (16.4% – 21.0%), but ductility values are essentially unaffected

Series I – Series III: Comparison is more qualitative because of the different sheeting system. There are no differences as fare as initial rigidity is concerned; however an increase of ductility has been expected. This was not possible as failure mode for the strap-braced specimens was not the most advantageous one. Strap braced wall panels have the advantage of stable hysteretic loops, but also the disadvantage of higher pinching than the sheeted ones.

Series I – Series OSB I: Comparison more qualitative, keeping in mind the different wall panel arrangements. Initial rigidity is of similar magnitude, with increase of ultimate load. Failure of OSB specimens under cyclic loading was more sudden than in case of corrugated sheet specimens, where degradation occurs gradually. This is also reflected by the reduced ductility for OSB specimens.

Series OSB I – Series OSB II: The effect of opening produced similar results as in cases of Series I – Series IV. Initial rigidity decreased with 64.6% – 59.1%, while ultimate load decreased with 32.5% - 36.9%. There is also an important decrease of ductility, probably highlighting the different failure modes of the two wall panels.

Performance criteria. In case of wall panels clad with corrugated sheeting damage is largely concentrated in seam fasteners. If plasticization in vicinity of fasteners increases the waterproof cladding layer looses its functionality and has to be replaced. For establishing global performance criteria the following acceptable deformations in the seam fasteners have been considered:

- If slip of the seams does not exceed the elastic limit, corresponding to $0.6F_{max}$ of the seam connection, damage is limited and can be considered negligible. In this case the cladding is still water-proof, no repairs are required and this would correspond to normal serviceability conditions.

- If slip is limited to the diameter of the screw (4.8mm) the cladding requires repair. There is damage, but not excessive and by minor interventions, like replacing screws with larger diameter ones the structure can be repaired. This could correspond to immediate occupancy criteria.

- In case of life safety criteria any kind of damage is acceptable, without endangering the safety of occupants. This criteria is not any more related to serviceability, but can correspond to the attainment of the ultimate force (F_{ult}) of the wall panel and the starting of the downwards slope.

Relative slip in seams has been measured for specimens I-3, II-2 and II-3 the first two criteria can easily be applied and relationship between slip and lateral deformation of the panel can be found. Based on these assumptions the following performance criteria are suggested for wall panels clad with corrugated sheet: (1) fully operational ($\delta<0.003$); (2) partially operational ($\delta<0.015$); (3) safe but extensive repairs required ($\delta<0.025$). Comparable design criteria can be established for other types of panels. The first performance level does not provide ductility, because shear panel work is limited to elastic domain. This could be the design criteria for frequent, but low intensity earthquakes. In case of rare but severe earthquakes, the last two design criteria can be used and some ductility will be available.

Simplified Finite Element Model for 3D Analysis of whole framing structure. In order to be able to provide suitable tool for time-history analysis of entire building structure the global behaviour of the wall panel has to be caught. A very detailed modelling of the panel, based on individual connection behaviour could not be sufficiently simple to for this purpose, so modelling had to be simple and efficient on one hand, and accurate on the other. The basic idea is to replace the shear panel with equivalent cross bracing system, a technique often used in elastic calculations for design purposes. A FE model based on DRAIN-3DX (Prakash and Powell, 1993) computer code is proposed in order to get as accurate hysteretic behaviour as possible (Fulop, 2002). The main features of typical experimental hysteretic curve is presented in Figure 36, emphasising all the important factors affecting earthquake performance, namely load bearing capacity, ductility and pinching. The model consists of a mechanism frame and a special bracing. As all column ends are hinged the frame itself is a mechanism and it does not contribute to load bearing capacity. Braces are modelled as 'TYPE 8' fiber hinge (FH) beam-column elements with FH to accommodate the hysteretic behaviour (Figure 37). In order to calibrate the FE model experimental results from a full scale testing program have been used for comparison (Table 11).

Conclusions. Both test and numerical results demonstrated a good behaviour of panel under cyclic loading. Practically, the ductility of tested panels can be characterised by behaviour factor q of 2.2 to 2.6, which, obviously, are greater than the 1.5 value proposed in the latest draft of EUROCODE 8 (pr. EN 1998, draft 2001) for non-dissipative structures. Thus, even the framing structure itself doesn't posses ductility, the combination with shear walls has a low to moderate ductility which can be considered in seismic design.

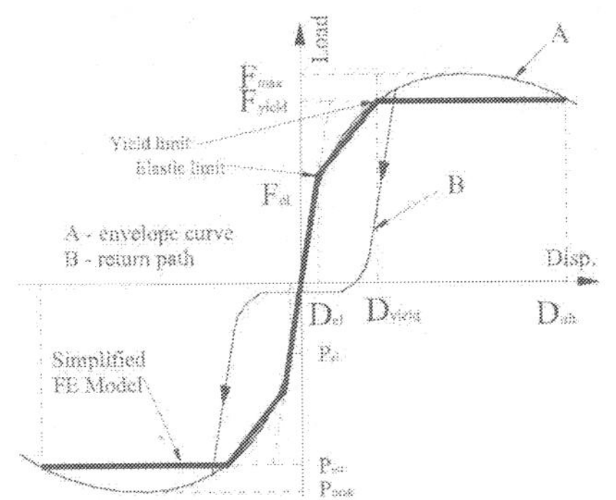

Figure 36. Scheme of hysteretic behaviour

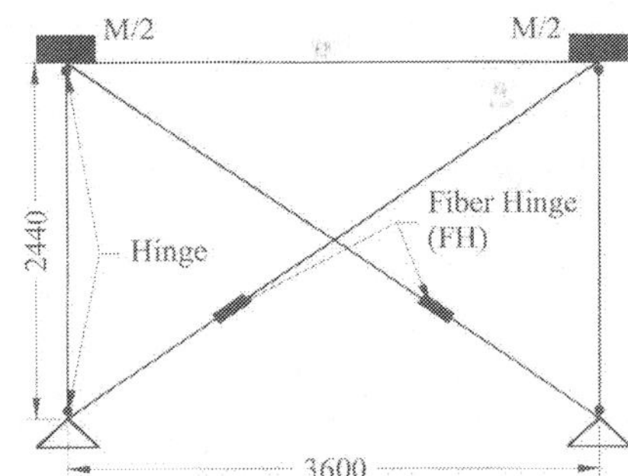

Figure 37. Simplified DRAIN-3DX model

Table 11. Series of FE models and calibrated values based on experimental results

Series	I	II	IV	OSB I	OSB II
Sheeting	Corrugated Sheet	Corr. Sheet + Gypsum	Corrugated Sheet	OSB	OSB
Initial Rigidity (N/mm)	3446.6	3850.6	1766.3	4197.3	1610.5
Elastic Limit (F_{el}/D_{el})(N/ mm)	24086/*6.99*	26566/*6.90*	128670/*7.28*	28942/*6.89*	11850/*7.36*
Yield Limit (F_{yield}/D_{yield})(N/ mm)	33560/*14.95*	39819/*15.58*	26812/*23.78*	48944/*17.49*	33908/*27.76*
Ultimate Limit (D_{ult}) (mm)	42.61	57.29	94.35	42.85	65.57

8.6.7 Conclusions

Cold-formed steel structures are non- or low-dissipative structures. They can be designed to resist seismic actions using an elastic design with a behaviour factor q equal to 1.5-2, as EN 1998-1 (2001) proposes. However, for special cases, and using an adequate design methodology (e.g. local plastic mechanism method, displacement based analysis or performance load criteria), a more optimistic design is possible. On this purpose, Eqn. 8 may be used.

But, due to the fact the seismic performance of such a type of structures is primarily based on redundancy and overstrength, the structural design criteria for structure and detailing, as they are stated in EN 1998, must be applied.

Therefore, due to the lightness of cold-formed steel structures, and if the conceptual design rules for seismic resistant structures are correctly applied, they are generally designed from the combination of permanent and variable actions, and not from the accidental ones, like the seismic action is; exception can be for bracing system, for some structural connections and for supporting details (Dubina et. al., 2000, 2001).

References

AISI (1997). Monotonic Tests of Cold Formed Shear Walls with Openings, Prepared by AHB research Center. Inc., The American Iron and Steel Institute.

AISI (1998). Shear Wall Design Guide, Publication RG-9804, AISI.

AISI- American Iron and Steel Institute (2001). North American Specification for the Design of Cold-Formed Steel Structural Members with Commentary, Washington, DC.

Bertero V.V. (1997). Earthquake Engineering. *Structural Engineering Slide Library (Golden W.G. Ed.)*, The University of Berkley. California, USA.

Bertero R.D., Bertero V.V. (1999). Redundancy in Earthquake- resistant Design. *Journal of Structural Engineering,* ASCE, Vol. 125 (1), 81-88.

Calderoni B., De Martino A., Landolfo R., Ghersi A. (1994) On the seismic resistance of light gauge steel frames. *Behaviour of Steel Structures in Seismic Areas STESSA '94.* Ed. by F.M. Mazzolani and V. Gioncu, E&FN Spon, London, ISBN 0-419-19890-3.

Calderoni B., De Martino A., Ghersi A., Landolfo R. (1997) Influence of local buckling on the global seismic performance of light gauge portal frames. *Behaviour of Steel Structures in*

Seismic Areas STESSA'97, Ed. by F.M. Mazzolani and Hyroshi Akiyama, Edizioni 10/17, Salerno, ISBN 88-85651-76-3.

De Martino A., Ghersi A., Landolfo R., Mazzolani F.M. (1992). Calibration of a bending model for cold-formed section. *XIth International Specialty Conference on Cold-Formed Steel Structures*, St. Louis, 503-511.

Dubina D., Fülöp L., Ungureanu V., Nagy Zs. (2000) Cold-formed Steel Structures for Single Storey Buildings. *International Conference on Steel Structures of the 2000's*, 11-13 September 2000, Istanbul, Turkey, 191-196.

Dubina D., Georgescu M., Ungureanu V., Dinu F. Innovative cold-formed steel structures for one storey penthouse superstructure of DATATIM-ALCATEL industrial building. *The 9th Int. Conference on Metal Structures – ICMS'2000*, Timisoara, Romania, 19-22 October 2000, 318-326.

Dubina D., Ungureanu V. (2000). Elastic-plastic intercative buckling of thin-walled steel compression members. *Proceedings of the 15th International Specialty Conference on Cold-Formed Steel Structures,* Oct. 19-20, 2000, St. Louis, Missouri, USA, 223-237.

Dubina D., Ungureanu V., Georgescu M., Fülöp L. (2001): Innovative cold-formed steel structure for restructuring of existing RC or masonry buildings by vertical addition of supplementary storey. *The 3rd International Conference on Thin-Walled Structures*, Cracovia, Poland, 5-7 June 2001, 187-194.

Dubina D., Ungureanu V., Fülöp L., Nagy Zs., Larsson H. (2001): LINDAB Cold-Formed Steel Structures for Small and Medium Size Non-Residential Buildings in Seismic Zones. *The 9th Nordic Steel Construction Conference – NSCC2001*, Helsinki, Finland, 18-20 June 2001, 463-470.

Dubina D., Ungureanu V., Nagy Z. (2002). Lightweight steel structures using Lindab cold-formed sections for residential and non-residential buildings, *International IASS Symposium on Lightweight Structures in Civil Engineering*,Warsaw, Poland, 24-28 June, 2002.

ECCS (1985) Recommended Testing Procedure for Assessing the Behaviour of Structural Steel Elements under Cyclic Loads.

ECCS (1987) European Recommendations for the Design of light gauge steel members. ECCS, Brussels, 1987.

ECCS (1995) European Recommendations for the Application of Metal Sheeting acting as a diaphragm, 1995.

ECCS (1995) European Recommendations for the Application of Metal Sheeting acting as a diaphragm. Pub.88, European Convention for Constructional Steelwork, Brussels.

EUROCODE 3 Part 1.3 (1993) Design of Steel Structures. Part 1.3: Cold-formed Thin Gauge *Members and Sheeting, European Committee for Standardisation*, 1993.

EUROCODE 3 (2001) EN 1993-1.1 Design of Steel Structures. Part 1.1: General Rules and Rules for Buildings, *European Committee for Standardisation*, Final Draft.

EUROCODE 3 (2001) pr. EN 1993-1.8 Design of steel structures. Part 1.8: Design of joints.

EN 1993-1-3 EUROCODE 3 (Draft of June 2001). Design of Steel Structures, Part 1.3: General Rules, Supplementary Rules for Cold-Formed Thin-Gauge Members and Sheeting, CEN/TC 250/SC3 - *European Committee for Standardisation*, Brussels.

EUROCODE 8 (1998). EN 1998 Design provisions for earthquake resistance of structures, Final Draft.

Fajfar P. (2000) "General definitions and basic relations" Chapter 8.1 in the book "*Moment Resistant Connections of Steel Frames in Seismic Areas. Design and Reliability*" edited by F.M. Mazzolani, E&FN Spon, London, 2000.

FEMA-273 (1997). NEHRP Guidelines for the Seismic Rehabilitation of Buildings, Prepared by the Building Seismic Safety Council for the Federal Emergency Management Agency, Washington.

Fulop L. (2002) Calibration of an equivalent bracing system for 3D dynamic inelastic analysis of steel framed building structures with dissipative shear walls. *International Colloquium Stability and Ductility of Steel Structures*, Hungary, Budapest, September 26-28, 2002.

Fulop L., D. Dubina (2002) Performance of wall stud shear walls under monotonic and cyclic loading. *International Colloquium Stability and Ductility of Steel Structures*, Hungary, Budapest, September 26-28, 2002.

Gad E.F., Chandler A.M., C.F. Duffield, G. Hutchinson. (1999). Earthquake Ductility and Overstrength in Residential Structures, Structural Engineering and Mechanics, 8:4, 361-382

Gad E.F., Duffield C.F. (2000) Lateral Behaviour of Light Framed Walls in Residential structures, *12 th World Conference on Earthquake Engineering.*

Ghobarah A. (2001). Performance based design in eartquake engineering: state of development. *Engineering Structures 23*, Elsevier, 2001, 878-884.

Gianfranco de Matteis. (1998) The Effect of Cladding in Steel Buildings under Seismic Actions, *PhD Thesis*, Universita degli Studi di Napoli Federico II.

Gianfranco de Matteis, Landolfo R., Mazzolani F.M., Fulop L., Dubina D. (1999). Seismic response of MR steel frames with different connection behaviours, *Proceedings of the 6th International Colloquium SDSS'99*, Elsevier, 409-420.

Gioncu V., Mazzolani F. M. (2002) Ductility of seismic resistant structures. *Spon Press*, London, 2002, ISBN 0-419-22550-1.

Johnston B.G. (1971) Spaced Steel Columns. *Journal of the Structural Division, ASCE*, Vol. 97, No ST5, p. 1465-1479.

Kawai Y., Kanno R., Hanya K. (1997) Cyclic Shear Resistance of Light-Gauge Steel Framed Walls, ASCE Structures Congress, Poland

Kawai Y., Kanno R., Uno N., Sakumoto Y. (1999) Seismic resistance and design of steel framed houses, *Nippon Steel Technical report*, No. 79.

Lim J. (2001) Joint effects in cold-formed steel portal frames. *PhD Thesis,* University of Nottingham, 2001.

Mazzolani F.M., Piluso V. (1996). Seismic design of resistant steel frames, E&FN Spon, London, 1996.

Moldovan A., Petcu D., Gioncu V. (1999) Ductility of thin-walled members. *Stability and Ductility of Steel Structures SDSS'99*. Ed. by D. Dubina and M. Ivanyi, Elsevier Science, 1999.

Murray N.W., Khoo P.S., (1981) Some basic mechanism in the local buckling of thin-walled steel structures. *International Journal of Mechanical Sciences*, Vol. 23, No. 12, 1981, 703-713.

Niazi A. (1993). Contribution a l`etude de la stabilite des structures composees de profils a parois minces et section ouverte de type C. *PhD Thesis*, University of Liege, Belgium.

Ono T., Suzuki T. (1986) Inelastic behaviour and earthquake - resistance design method for thin-walled metal structures. *Proceedings on IABSE Coll. On Thin-Walled Metal Structures in Building*, Stockolm, 115-122.

Paulay T., Priestley M.J.N. (1992) Seismic Design of Reinforced Concrete and Mansonry Buildings, *J. Wiley and Sons*, 1992.

Prakash V., Powell G.H.. (1994). Drain-3DX Base program description and user guide, Version 1.10, Department of Civil Engineering, University of California at Berkley.

Rondal J., Niazi M (1990) Stability of built-up beams and columns with thin-walled members. *International Colloquium Stability of Steel Structures*, Budapest, Hungary.

Salenicovich A.J.et al. (2000), Racking Performance of Long Steel-Frame Shear Walls, *Fifteenth Int. Speciality Conference on Cold-Formed Steel Structures*, St. Louis, Missouri, Oct. 19-20, 471-480.

Serrette R.L., Hall G., Nygen J. (1996). Shear Wall Values for Light Weight Steel Framing, AISI.

Serrette R.L. (1998). Seismic Design of Light Gauge Steel Structures: A discussion, Fourteenth *Int. Speciality Conference on Cold-Formed Steel Structures*, St. Louis, Missouri, Oct.15-16

Schafer B.W., Pekoz T. (1998). Computational modelling of cold-formed steel characterising geometric imperfections and residual stresses, *Journal of Constructional Steel Research* No. 47 (3) P. 193-210.

Serette R.L., Ogunfunmi K. (1996). Shear Resistance of Gypsum-Sheeted Light-gauge Steel Stud Walls, *Journal of Structural Engineering*, ASCE, 122:4.

Wittaker, G. Hart, C. Rojahn. (1999). Seismic response modification factors, *Journal of Structural Engineering* 1999:2, 438-443.

Wilkinson T. (1999). The plastic behaviour of cold-formed rectangular hollow sections, *PhD Thesis*, The University of Sydney, Australia, 1999.

Wong M.F., Chung K.F. (2002) Structural behaviour of bolted moment connections in cold-formed steel beam-column sub-frames. *Journal of Constructional Steel Research* 58 (2002) 253-274.

Zadanfarrokh F., Bryan E. R. (1992) Testing and design of bolted connections in cold-formed steel sections, *Eleventh International Specialty Conference on Cold-Formed Steel Structures*, St. Louis, Missouri, USA, October 20-21, 1992.

Zaharia R. (2000) Contributions to the safety study of the cold-formed steel structures, *PhD Thesis*, The 'Politehnica' University of Timisoara, 2000.

Zaharia R., Dubina D. (2000) Behaviour of cold-formed steel truss bolted joints, *The IV'th International Workshop on Connections in Steel Structures*, Roanoke, USA, October 2000.

Chapter 9: Pallet Racking

J. M. Davies

The Manchester School of Engineering, University of Manchester, Manchester, England
E-mail: jmdavies@fs1.eng.man.ac.uk

9.1 Introduction

Nowadays, the storage industry is huge and, in the UK alone, about 50000 tonnes of steel are used annually in the fabrication of structures whose sole purpose is to store material in a manner that makes it easily accessible (Godley; 1991). Much of this steel is in the form of cold-formed sections so that storage structures are a particularly important application of light gauge steel technology. As a consequence of this, the storage industry has often led the way with research to improve the state of the art and thus to define better design procedures for cold-formed steel members. Furthermore, the design of pallet racking is particularly demanding and offers insights into many aspects of the practical design of thin-walled structures. It is, therefore, particularly appropriate to study pallet racking as part of this series of papers. In producing this paper, a few minor simplifications have been made in order to introduce the essentials of the subject without an over-concentration on points of detail. Accordingly, this paper should be seen as offering an introduction to the subject rather than a detailed design procedure.

There are many different types of storage structure, but, for general purposes, it is possible to divide them, in increasing order of size, as:

- Shelving systems
- Pallet racking
- Rack buildings

The term "shelving" is self explanatory although shelving systems can be very sophisticated with computerised inventories and automatic retrieval systems. Pallet racking is the "work horse" of the storage industry and allows goods of all descriptions to be stored on standard-sized pallets. The basic system is typically about 6 metres high and is serviced by manually-operated fork lift trucks. However, high bay pallet racking can be up to 30 or more metres high serviced by computer-controlled mechanical handling equipment. In a rack building, high bay pallet racking not only provides the storage system but also forms the primary structure of the building. Rack buildings can be huge, 30 m high × 30 m wide × 100 m long being quite typical. Because of their significance in the context of the global usage of cold-formed steel sections, most books on cold-formed section design include a chapter on storage racking (Godley, 1991; Hancock, 1998; Hancock et al., 2001).

This paper is concerned with typical pallet racking, as illustrated by Fig. 1. The primary features of the rack structures considered in this paper, and these are typical of the European industry, are the exclusive use of cold-formed sections for all of the members and the use of clipped connections between the uprights and the columns. The clipped connections allow beams to positioned at various levels to suit the characteristics of the goods being stored and result in the beam to column connections having a semi-rigid performance and this dominates the structural design. In order to accommodate the requirement of flexibility in the positioning of the beams, the uprights are perforated with a regular array of holes and/or slots into which the beam end connectors are slotted. Significantly, from the point of view of design, pallet rack structures may be either braced or unbraced in the down-aisle direction as shown in Figs. 2 and 3 respectively.

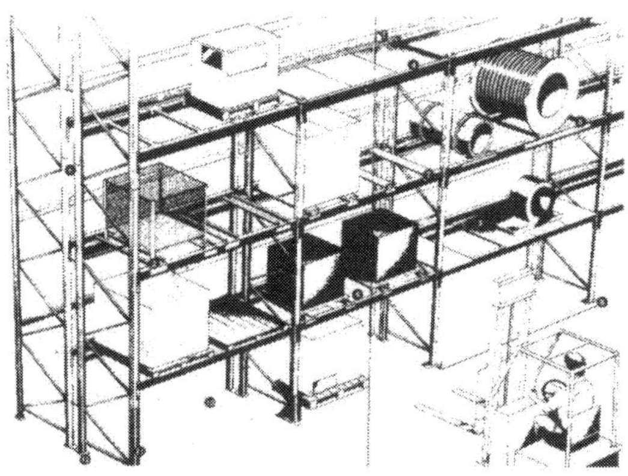

Figure 1. Typical pallet racking

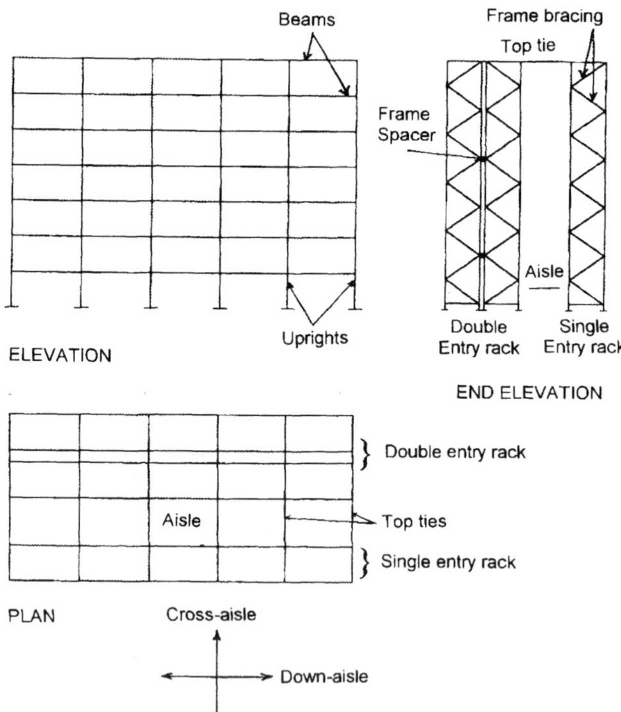

Figure 2. Typical configuration of an unbraced pallet rack structure

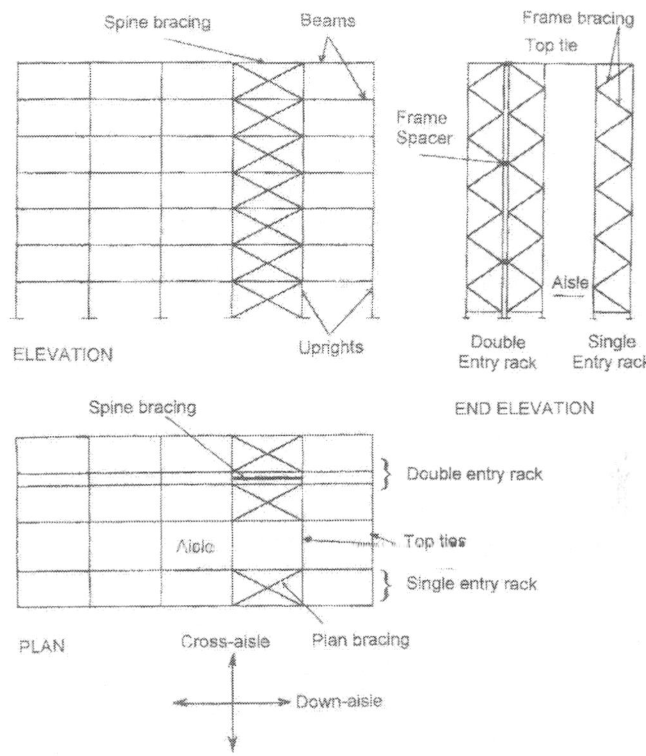

Figure 3. Typical configuration of a braced pallet rack structure

It follows that, for the purposes of structural design, the significant characteristics of pallet rack structures are:

- Specific load cases which take into account the usage
- The necessity for second-order structural analysis with semi-rigid joints, bearing in mind the likelihood of a low elastic critical load
- The use of perforated uprights of open cross-section (usually singly symmetrical)
- The performance of semi-rigid clipped connections
- The performance of semi-rigid bases to the uprights
- The use of testing where necessary and the appropriate test methods

These characteristics will be explored in more detail in the sections which follow. For simplicity, attention will be concentrated on unbraced racks which include all of the essential characteristics. Braced racks are inherently more stable in the down-aisle direction. The only additional factor in their design is the eccentricity of the down-aisle bracing system. Similarly, there are a number of other sub-types such as drive-in and drive-through racks which are not considered here.

For the purposes of global structural design, a pallet rack is similar to a multi-storey building. Because it is of very slender construction with semi-rigid joints, and because it is likely to be fully loaded a number of times in its lifetime, the design is particularly interesting and demanding.

9.2 Design standards for pallet racking

The design of pallet racking is strongly influenced by the available design standards for cold-formed sections in general. Thus, the two main standards internationally are, in Europe, the design standard of the Federation Europeenne de le Manutention (FEM, 1999) and, in the USA, the design specification of the Rack Manufacturers Institute (RMI, 1997). The FEM standard is strongly dependant on Part 1.3 of Eurocode 3 (EC3, 1996 which will be referred to as EC3) while the RMI specification is equally dependant on the American standard for the design of cold-formed steel structural members (AISI, 1999).

The racking standards themselves are concerned with the issues that are specific to pallet racks and these are summarised in section 1 above. It is of note that the two international standards are about as different as two standards concerned with the same subject could be (Davies and Godley, 1998). These notes will concentrate on the European design procedure and will make frequent reference to the FEM code.

9.3 Design loads and load combinations

Evidently, the primary loads are those arising from the pallets themselves and, in the FEM code, these have a load factor of $\gamma_Q = 1.4$. There are also the "placement loads" arising from any impact effects during handling. Vertical placement loads only affect the design of the beams themselves, but horizontal placement loads, in both the down-aisle and cross-aisle directions have a significant effect on the global design. The value of the horizontal placement load depends on the height of the rack and how the material is handled and a value of 0.5 kN with a load factor of $\gamma_Q = 1.4$ is typical.

"Frame imperfections", which are related to the out of plumb of the uprights and any initial connector looseness, are also important. The sway imperfection φ, also applied with $\gamma_Q = 1.4$, is particularly significant and is given by:

$$\varphi = \sqrt{\left(\frac{1}{2}+\frac{1}{n_c}\right)}\sqrt{\left(\frac{1}{5}+\frac{1}{n_s}\right)}\left(2\varphi_s+\varphi_\ell\right) \leq \left(2\varphi_s+\varphi_\ell\right)$$

where n_c = number of uprights in the down-aisle direction or number of frames in the cross-aisle direction

n_s = number of beam levels

φ_s = maximum specified out of plumb divided by the height

φ_ℓ = looseness of the beam to upright connector (see later)

These sway imperfections apply in both horizontal directions but need only be considered in one direction at a time. They are generally replaced by a closed system of horizontal forces applied at each beam level as shown in Fig. 4

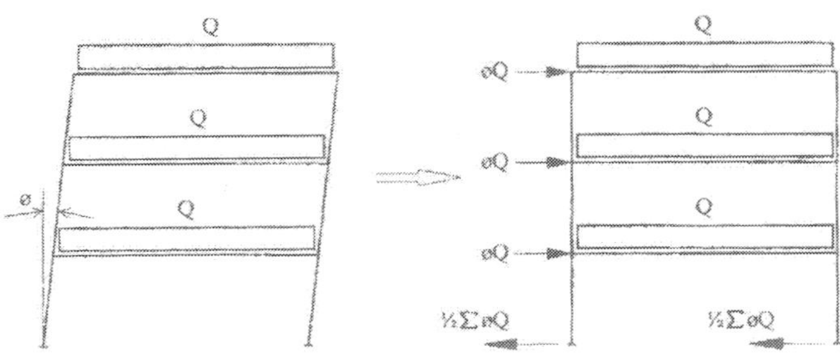

Figure 4. Sway imperfections and equivalent horizontal forces

Finally, the design of the uprights is strongly influenced by pattern loading. After investigating a number of potentially critical patterns, it has been concluded that, for the design of the critical uprights in the lower storeys of the rack, it is sufficient to consider one or other of the patterns shown in Fig. 5 (depending on whether the lowest beam is close to the ground).

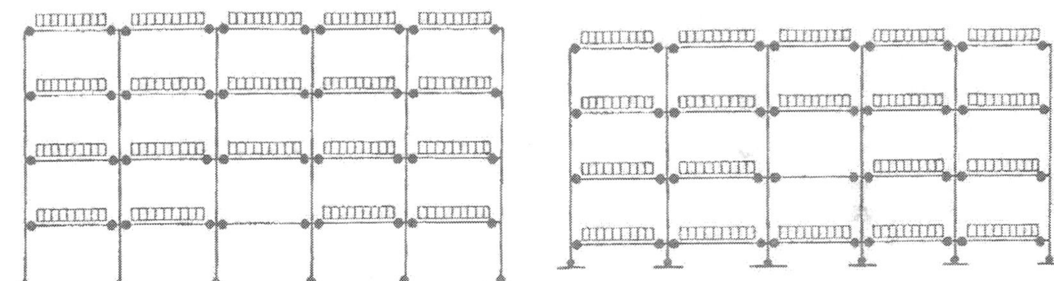

Figure 5. Critical pattern loads for upright design

It follows that the critical load combinations are as follows bearing in mind that, when all unfavourable actions occur simultaneously, a combination factor of 0.9 is allowed:

<u>Down-aisle analyses</u>

D1	Pallet loads only with γ_Q
D2	Pallet loads and imperfections with γ_Q
D3	Pallet loads, imperfections and placement loads all with $0.9\gamma_Q$

Pattern loading is considered separately and the additional bending moments added to D1, D2 or D3 as appropriate.

Cross-aisle analyses

C1 Vertical loads only with γ_Q (causes no bending moments)
C2 Vertical loads and imperfections with γ_Q
C3 Vertical loads, imperfections and placement load (P1) all with $0.9\gamma_Q$
C4 Vertical loads, imperfections and placement load (P2) all with $0.9\gamma_Q$

Note: For overall stability, the most unfavourable position of the placement load is at the level of the top beam (P1). In principle, the horizontal placement load can arise at any beam level and the nodes of the upright frames may not coincide with the beam levels. In general, therefore, there is a second placement load case to consider (P2) which gives rise to the maximum bending moment in the upright. This does not require a global analysis and, for convenience, this load case is termed C4.

Combined down-aisle and cross-aisle analyses

The FEM code specifies that both imperfections and placement loads act only in one direction at a time and that it is not necessary to consider the interaction of down-aisle imperfections with cross-aisle placement loads and vice versa. This leads to the following relatively simple load combination cases in the design of the uprights, all of which include pattern loading in the down-aisle direction:

1) D1 + C2
2) 0.9D1 + worst of C3 or C4
3) C1 + D2
4) 0.9C1 + D3

9.4 Cold-formed steel members and connections

The primary structural elements of a pallet rack are the uprights, the beams and the bracing members. All are generally cold-formed sections. Some typical upright sections are shown in Fig. 6. The uprights are generally mono-symmetric sections which have regular arrays of perforations to receive the beam end connectors. The design of these uprights is critical. There are a number of possible beam configurations and Fig. 7 shows some of the most common.

A fundamental feature of pallet racking is the use of clipped beam to upright connections as shown in Fig. 8. These connections invariably have a semi-rigid moment-rotation characteristic that dominates the design procedure. A similar situation arises at the base of the uprights where a proprietary baseplate arrangement, such as is shown in Fig. 9, provides a semi-rigid connection to the floor. The rotational stiffness of both of these connections has to be determined by test. It is, of course, safe to ignore the stiffness of the upright to floor connection and to assume a pinned joint.

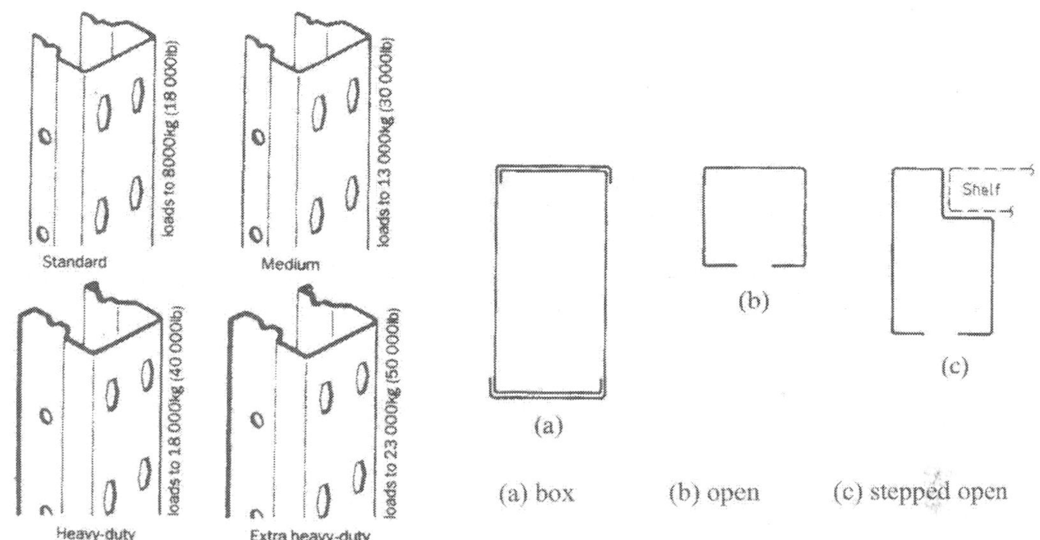

Figure 6. Typical upright profiles (Godley, 1991)

Figure 7. Typical beam sections (Godley, 1991)

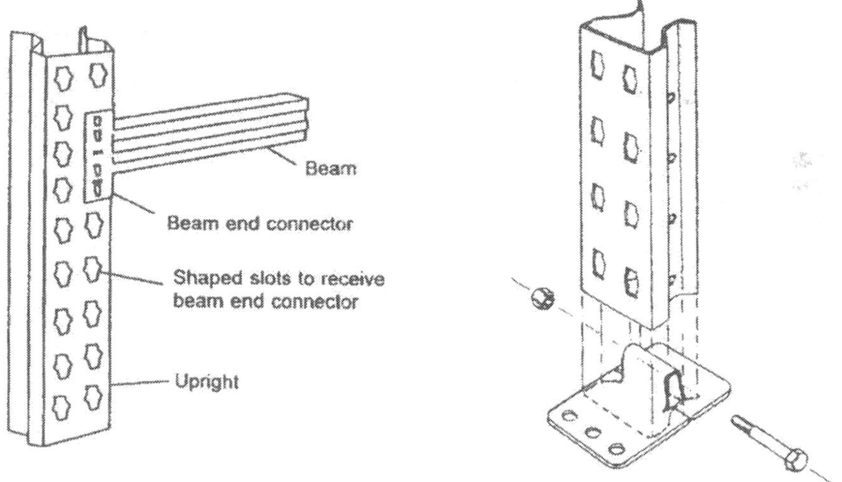

Figure 8. Semi-rigid beam to upright connection

Figure 9. Semi-rigid baseplate connection

The FEM code describes a suitable test for the determination of the mechanical properties of the baseplate connection at the foot of an upright. However, these properties are dependent on the axial load so that it is necessary to carry out the test for a range of axial loads. As it is possible to assume a pinned connection, for simplicity, this aspect of the design will not be pursued further here.

9.5 Global analysis: second-order analysis with semi-rigid joints

The design of a pallet rack system is verified in two stages. In the first stage, a global analysis of the structure must be made in order to verify the overall stability and to determine the distribution of internal forces and displacements. In the second stage, the individual elements of the structure must be checked to ensure that they have adequate resistance in the ultimate limit state and that unacceptable deformations do not develop in the serviceability limit state. For the purposes of global structural analysis, the two orthogonal directions, termed "down-aisle" and "cross-aisle", may be considered separately and their stress-resultants added for the purposes of member design.

Pallet racks are relatively slender structures with semi-rigid joints. It follows that they are extremely sensitive to second-order (PΔ) effects. Indeed, when analysing an unbraced rack in the down-aisle direction, it is not unusual to find that the elastic critical load is not significantly greater than the design load. This aspect of the design has, therefore, to be treated seriously and some special analysis techniques are available for investigating down-aisle stability. In general, in each of these methods, the semi-rigid beam to upright connections and the upright base connections are assumed to have a linear moment-rotation characteristic up to their design strength. However, the use of a multi-linear curve is not precluded. The experimental determination of the mechanical properties of the beam to upright connection is discussed in the next section.

In the FEM code, following closely principles that are used in EC3, the appropriate analysis is determined on the basis of "frame classification" which is based on the elastic critical load ratio V_{Sd}/V_{cr} where:

V_{Sd}	is the design value of the vertical load on the frame
V_{cr}	is the elastic critical value of the vertical load for failure in a sway mode

a) If $V_{Sd}/V_{cr} \leq 0.1$, the frame is classified as non-sway and a first-order analysis is sufficient.

b) If $0.1 < V_{Sd}/V_{cr} \leq 0.3$, a <u>level 2</u> analysis may be used in which second-order effects are treated indirectly.

c) If $V_{Sd}/V_{cr} > 0.3$, a <u>level 1</u> analysis is required in which second-order effects are treated directly.

Clearly, an estimate of the elastic critical load is fundamental to pallet rack design and the following level 2 methods are available and have sufficient accuracy. In each case, they also include a level 2 estimate of the enhancement of the bending moments due to second-order effects:

1. The "Horne" or "Amplified sway" method may be used whereby the elastic critical load is estimated on the basis of a first-order sway analysis. This method is surprisingly accurate and its basis is given in (Horne, 1975), it is also included in the main part of Eurocode 3 (ENV 1993-1-1).

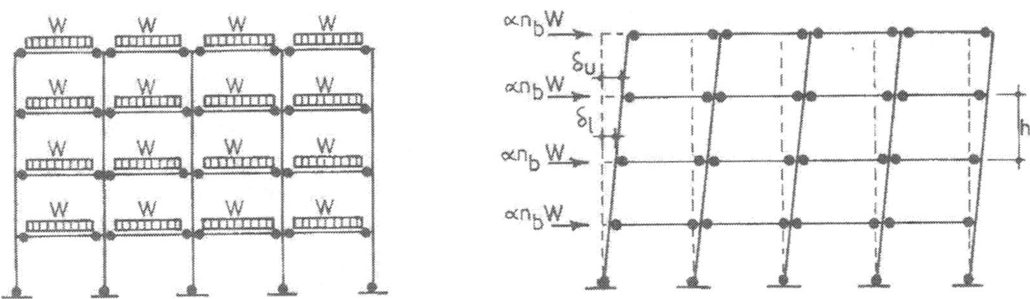

Figure 10. The "Horne" method of estimating the elastic critical load V_{cr}

The first part of Fig. 10 shows the actual frame and the vertical pallet loading and the second part shows these vertical loads changed into notional horizontal loads and the resulting deflections where φ is an arbitrary multiplier (generally taken as the sway imperfection factor) and n_b is the number of bays. The elastic critical load V_{cr} is then given by

$$\frac{V_{cr}}{V_{Sd}} = \frac{\phi}{\phi_{\max}}$$

where φ_{max} is the largest value of the sway index φ_s in any storey and

$$\phi_s = \frac{\delta_U - \delta_L}{h}$$

At the required limit state, the design values of the internal forces and deflections in any sway mode are those determined by a first-order elastic analysis amplified by the factor β where:

$$\beta = \frac{V_{cr}}{V_{cr} - V_{Sd}}$$

2. The explicit equations given in Appendix B of the FEM code may be used. These equations are based on using a "Grinter" substitute frame in order to simplify the Horne method. A further simplification then assumes that the global sway stability is dominated by the sway of the first two storeys. The result is a remarkably accurate series of equations that allow the elastic critical load and the sway amplification factor to be estimated "on the back of an envelope". The basis of these equations is given in Davies (1992).

3. In the cross-aisle direction, the upright frames are invariably braced. It is unusual for second-order effects to be significant although this needs to be checked and a simple estimate, based on smoothing out the shear flexibility of the bracing system, is given in Appendix C of the FEM code.

It may be noted here that, in general, and in contrast with more conventional structures using hot-rolled sections, it is not possible to assume that the joints of a cold-formed structure are rigid. The FEM procedure does not assume that the cross-aisle bracing system is rigid but, instead, determines its global stiffness by a test of the type shown in Fig. 11.

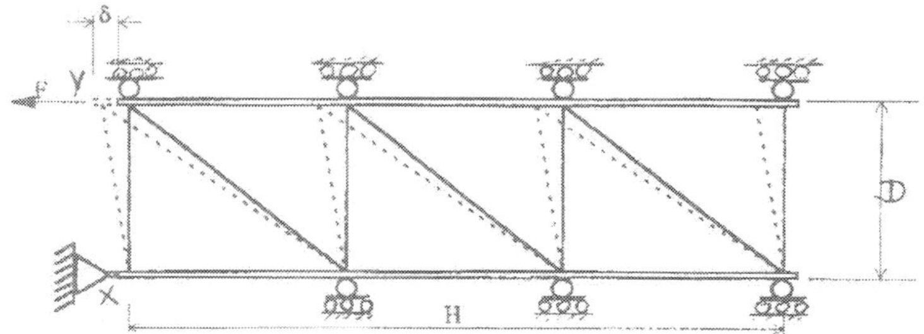

Figure 11. Determination of the shear stiffness of a braced upright frame

The bracing shear stiffness determined in this way may be used in two ways. It provides a necessary term in the equation for the approximate elastic critical load which is used in frame classification. It also allows an equivalent (reduced) cross sectional area to be calculated for the bracing members for use in the global analysis of the upright frames.

The outcome of the global analyses is sets of stress resultants (axial loads, bending moments and shear forces) and deflections for the various load cases in both the down-aisle and cross-aisle directions. These provide a global stability and serviceability check and also the information necessary to carry out the necessary checks on the components of the structure. The most important of the deflection limits are:

Maximum vertical deflection in a beam — span/200

Maximum sway at the top of the structure (defined as the movement in addition to any initial out of plumb not including placement loads) — height/200

9.6 Determination of the strength and stiffness of the beam to upright connection

The test arrangement for the determination of the properties of the beam end connector is shown in Fig. 12. This arrangement needs to be incorporated into a relatively stiff testing frame in such a way that a short length h_c of upright has no contact with the frame, where:

$$h_c > \text{beam connector length} + 2 \times \text{column face width.}$$

The test is carried out by first applying a small preload to remove any initial looseness (the "looseness" is measured separately) and then loading in increments to failure and the outcome is a graph of moment $M = aF$ versus $\theta = (\delta_2 - \delta_1) / d$ where:

a	= lever arm for the load F
d	= distance between the gauges C_1 and C_2
δ_1	= deflection measured by gauge C_1
δ_2	= deflection measured by gauge C_2

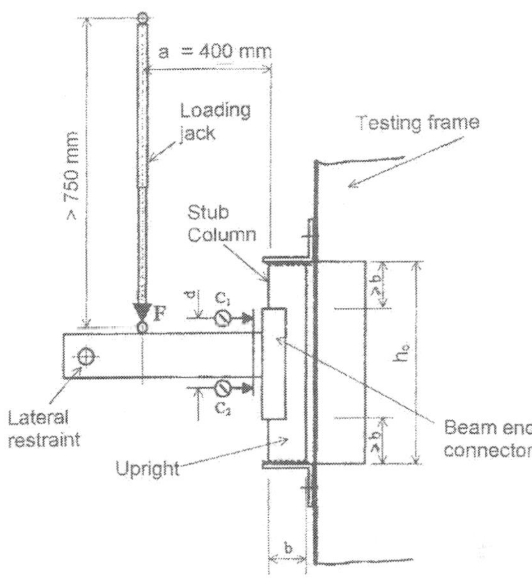

Figure 12. Apparatus for carrying out a test on a beam end connector

The interpretation of the results is shown in Fig. 13. M_{ti} is the maximum observed moment in test 'i'. The characteristic failure moment M_k is found by carrying out several such tests and interpreting the results statistically in order to obtain an acceptable level of reliability. The standard procedure for all situations of this nature is given here by:

$$M_k = M_m - ks$$

Where: M_m is the mean value of the maximum observed moments
s is the standard deviation
k is a statistical coefficient which reduces as the number of tests 'n' in the series increases as shown in Table1.

The design moment for the connection is then $M_{Rd} \leq \dfrac{M_k}{\gamma_M}$ where γ_M is the partial safety factor for connections (= 1.25).

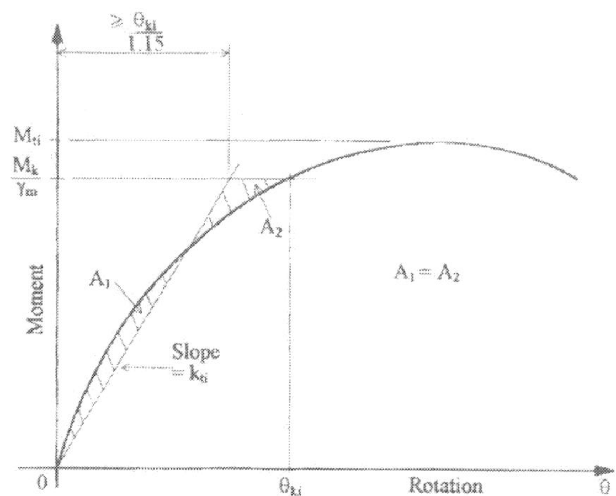

Figure 13. Interpretation of beam end connector tests

Table 1. Statistical coefficients used in the evaluation of test results

n	3	4	5	6	7	8	10	15	20	50	∞
k	3.15	2.68	2.46	2.33	2.25	2.19	2.10	1.99	1.93	1.81	1.64

The rotational stiffness of the connector, k_{ti}, is obtained as the slope of a line through the origin which isolates equal areas between it and the experimental curve below the design moment M_{Rd}.

The same test arrangement can be used to determine an initial looseness of the connection, φ_{li}, for use in determining the frame imperfection parameter. In this case, the loading jack in Fig. 12 needs to be double acting and capable of applying a load in the reverse direction. Fig. 14 shows the typical outcome from such a test as well as the interpretation in terms of the connector looseness. The required looseness parameter is the mean looseness determined by several such tests.

9.7 Design of the uprights

The uprights in pallet racks may be influenced by all three generic types of buckling, namely local, distortional and global (lateral torsional) and all three are explicitly considered in the FEM design procedure. Because of their typical proportions, distortional buckling tends to have particular importance. The uprights are generally subject to axial load and bending about both orthogonal axes. As is usual in such cases, these three conditions are considered separately and then combined in a relatively simple interaction equation.

A complete procedure for the design of singly-symmetric cold-formed steel upright is given in EC3. However, here, the uprights have a regular array of holes and this is not covered by the Eurocode or, indeed, any of the major cold-formed steel standards. Furthermore, when the column is part of an upright frame, with cross-bracing in one orthogonal plane and beams in the other, the boundary conditions are not at all clearly defined.

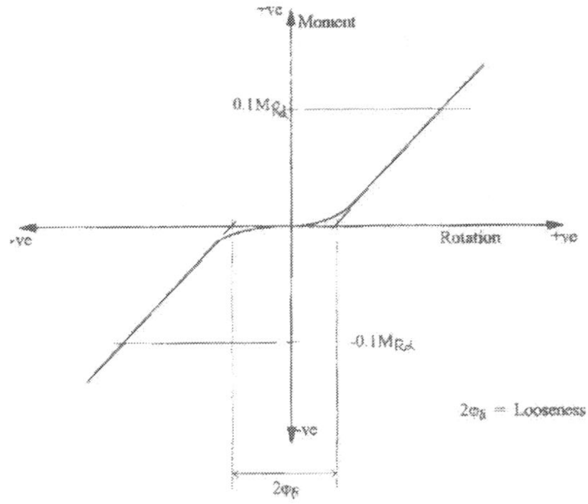

Figure 14. Interpretation of a connector looseness test

Historically, the design of these uprights with respect to axial load has usually been carried out by testing a range of isolated columns of different lengths in order to determine a "column curve" giving the design strength as a function of slenderness. The FEM code now allows three alternative procedures to arrive at a suitable column curve as follows:

9.7.1 Design by testing

The first essential is to carry out a stub column test, as illustrated in Fig. 15 in which a short length of upright is axially loaded to failure. This is also an essential part of the third alternative design procedure considered in section 7.3. Because this is a generic test for cold-formed section columns in general, and because it illustrates a number of significant points regarding testing, it will be considered in some detail.

The "buckling length" should be greater than three times the greatest flat width of the section (ignoring any intermediate stiffeners) and should include at least five pitches of the perforations. The intention here is to obtain a local buckling failure taking account of the perforations and clear of any influence of the end plates. The specimen should be short enough to avoid any influence of distortional or global buckling. The base and cap plates may be either bolted or welded to the ends of the stub upright.

The FEM code allows two alternative ways of applying the load and there is a long-running, and inconclusive, debate regarding which is to be preferred. In the first method, the axial load is applied through ball bearings which are located in indentations in the base plates at each end. Notionally, the position of these bearings should be at the centroid of the locally buckled section but the precise position of this is, of course, unknown. Therefore, the position of the bearing may be adjusted by trial and error until the maximum failure load is found.

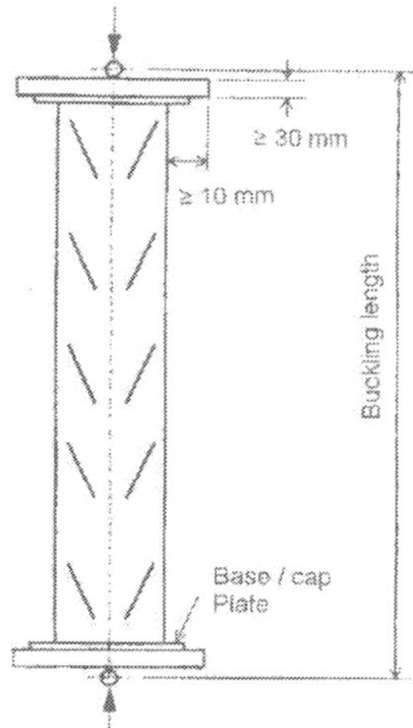

Figure 15. Arrangement for a stub column test

Alternatively, a testing machine may be used which allows one of the end loading platens to be adjusted by rotation about two horizontal axes and then clamped in position. There are no ball bearings in this procedure. Instead, the specimen is mounted in the testing machine with the centroid of its gross cross-section positioned centrally and a small holding load is applied in order to bring the adjustable loading platen just into full bearing on the end plates of the specimen. The adjustable platen is then clamped into position.

The load is then increased in increments up to failure and the maximum load carried is recorded.

When carrying out tests of this nature, it is inevitable that the actual thickness and yield stress of the test specimen do not coincide with the nominal values assumed for the purposes of design. It is, therefore, necessary to correct the test result to conform to the nominal values and the procedure for this is as follows:

$$R_{ni} = R_{ti}\left(\frac{f_y}{f_t}\right)^{\alpha}\left(\frac{t}{t_t}\right)^{\beta}$$

in which, for the test specimen:

R_{ni} = the corrected failure load for specimen 'i'
R_{ti} = the observed failure load for specimen 'i'
f_t = the observed yield stress for the specimen
f_y = the nominal yield stress of the steel
t_t = the observed thickness of the specimen

t	= the design thickness
α	= 0 when $f_y \geq f_t$
α	= 1 when $f_y < f_t$
β	= 1 unless certain conditions (specified in the code) prevail for $t < t_t$ and relatively slender plate elements when $\beta > 1.0$.

After a number of such tests have been carried out, the characteristic failure load, R_k is determined from

$$R_k = R_m - ks$$

Where: R_m is the mean value of the maximum observed failure loads
s is the standard deviation
k is a statistical coefficient which reduces as the number of tests increases

Having carried out the stub column tests, the design procedure continues by testing longer lengths of upright to failure in order to determine the complete column curve. In order to remove some of the uncertainty regarding the boundary conditions, the tests are carried out by testing an individual upright as part of a complete upright frame, as shown in Fig. 16. The frame width is the maximum used in practice and the bracing pattern, bracing sections and bracing connections are all the standard ones for the product.

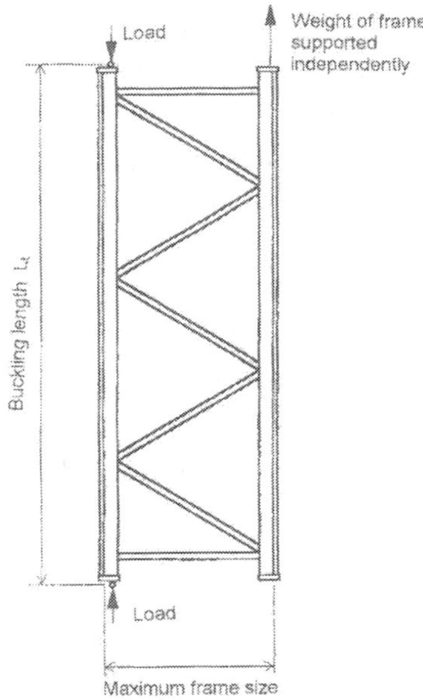

Figure 16. Compression test on an upright

The loaded upright is fitted with base and cap plates and loaded through ball bearings as discussed in connection with the stub column test above. The uprights are tested in a range of lengths, the smallest of which should just allow a single bracing panel. A total of at least five lengths should be tested in this way, the longest corresponding to a non-dimensional slenderness ratio:

$$\bar{\lambda} = \frac{\lambda}{\pi\sqrt{\frac{E}{f_y}}} = 1.50$$

for down-aisle buckling.

The test results are corrected to take account of the actual thickness and yield stress of the specimen in a similar manner to the stub column results above.

A statistically admissible (safe) column curve may then be derived as follows. For each test, including the stub column tests, the values of the stress reduction factor χ_{ni} and the non-dimensional slenderness ratio $\bar{\lambda}_{ni}$ should be calculated, where:

$$\chi_{ni} = \frac{R_{ni}}{A_g f_y}; \qquad \bar{\lambda}_{ni} = \frac{\lambda_{ni}}{\left(\frac{\pi^2 E}{f_y}\right)^{\frac{1}{2}}}$$

in which R_{ni} = adjusted failure load for test 'i'
A_g = gross cross-sectional area
f_y = nominal yield stress
λ_{ni} = slenderness ratio based on the gross cross section

A graph may now be plotted showing the values of χ_{ni} against $\bar{\lambda}_{ni}$. This is the initial column curve which must be corrected for the statistical variability of the test results. A suitable mathematical expression should first be chosen for χ_{cu} (subscript 'cu' for curve) to represent the locus of the mean values of the test results χ_{ni}. The choice of expression is entirely arbitrary except that it should be asymptotic from below to the elastic buckling curve $\chi = \frac{1}{\bar{\lambda}^2}$. Naturally, however, the more accurate the expression chosen to represent the experimental results, the more favourable will be the derived characteristic values.

The individual values of χ_{ni} are next normalised by dividing each one by the corresponding curve value, χ_{cu}. The standard deviation, s, of these normalised values may then be calculated.

Next, the characteristic value of the stress reduction factor, taking account of the statistical variation of the test results, should be calculated using:

$$\chi' = \chi_{cu}\left(1 - k\,s\right)$$

in which k is the statistical factor (≈ 2) given previously in Table 1 based on the total number of test results, including the stub column tests. It is important to note that this curve will not normally pass through unity at zero slenderness ratio because it is based on the gross cross-sectional area A_g. The effective cross-sectional area A_{eff} is not yet determined but is now simply given by:

$$A_{eff} = A_g\,\chi'_e$$

where χ'_e is the value of χ' at stub column slenderness.

Finally, the characteristic value of the stress reduction factor χ is given by:

$$\chi = \chi'\frac{A_g}{A_{eff}}$$

for values of $\bar{\lambda}$ given by:

$$\bar{\lambda} = \frac{\lambda_{ni}}{\left(\dfrac{\pi^2 E}{f_y}\right)^{\frac{1}{2}}}\left(\frac{A_{eff}}{A_g}\right)^{\frac{1}{2}}$$

which is the column curve to be used in design.

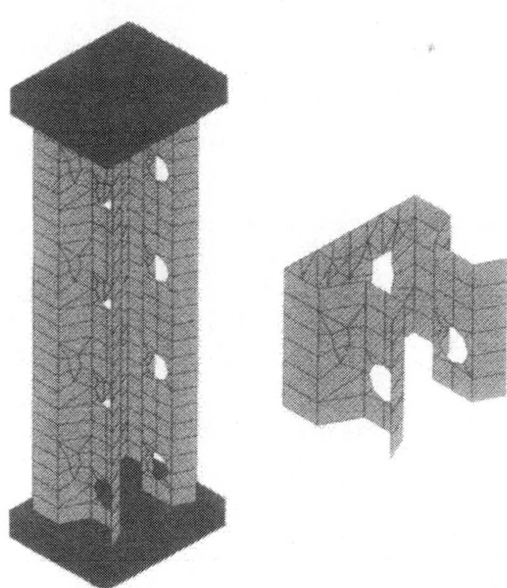

Figure 17. Finite element representation of a perforated upright

9.7.2 Full theoretical design procedure

The FEM code also allows the use of a fully theoretical design procedure for perforated compression members which takes rational account of the perforations together with local, global and distortional buckling and imperfections. It should be noted that, when designing perforated members by calculation, it is essential to consider the possibility of the local buckling of a strip of metal between two adjacent slots or large holes. Although the use of finite element techniques is not mandatory in this context, it is difficult to think of any other alternative. Fig. 17 shows a finite element representation of a perforated upright that has been used in research. As in the majority of such analyses, special care has to be taken in deciding on the appropriate boundary conditions.

9.7.3 Column curve based on stub column tests

The third alternative design procedure for the uprights of pallet racks uses the results of "stub column" tests, as shown in Fig. 15, as the basis for a theoretical derivation of the column design curve. If this procedure is used, it is necessary also to carry out a separate check to ensure that distortional buckling, as shown in Fig. 18, is not significant. If it is, a procedure is given whereby the column strength may be reduced to take account of distortional buckling

Figure 18. Distortional buckling in a racking upright

Having, therefore, carried out a series of stub column tests, as described above, a set of three distortional buckling tests, as shown in Fig. 18, are carried out on a single column of length equal to the length of a single bracing panel closest to 1 metre. The results of these tests are corrected for yield stress and thickness as before and the average failure load, $N_{db,Rd}$ determined.

The nominal strength, $N_{d,Rd}$, at this column length, in the absence of distortional buckling, is then calculated using the conventional procedures given in the appropriate design code (EC3) taking account of lateral-torsional buckling. If the ratio $N_{db.Rd}/N_{b.Rd}$ is less than unity, the characteristic stub column strength is then modified to take account of distortional buckling using:

$$N_{c.Rd} = \left(\frac{N_{db.Rd}}{N_{b.Rd}}\right)\frac{f_y A_{eff}}{\gamma_M}$$

The column curve may then be computed, on the basis of this stub column strength, in the usual way as described, for instance, in EC3.

9.7.4 Design of the uprights for axial load and bi-axial bending

The uprights in a pallet rack are subject to axial load together with bending moments in both the down-aisle and cross-aisle directions. The relevant stress resultants are available as a result of the global analyses considered above. Having determined the appropriate "column curve", the design then follows the relevant design equations for bending and compression with lateral torsional buckling, as given, for example, in EC3.

These interaction equations require the bending strength of the upright about both orthogonal axes. Because of the perforations, the required moments of resistance are best determined by simple four-point bending tests as shown in Fig. 19. The results of such bending tests are corrected for the thickness and yield stress of the material and interpreted statistically in the usual way.

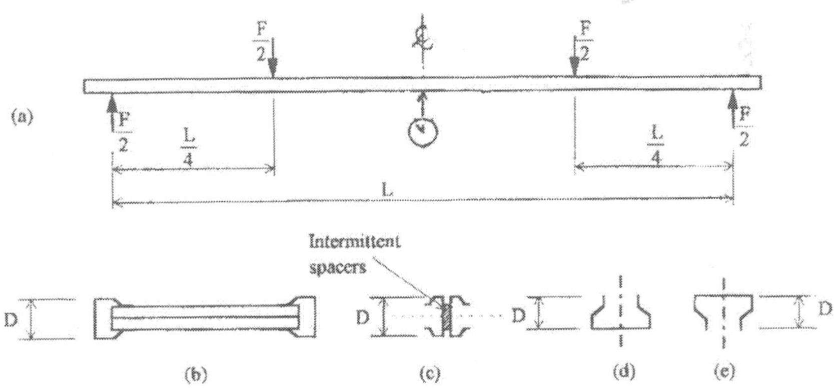

Figure 19. Bending tests on an upright section about the major and minor axes

Because of the number of load cases and the complex nature of some of the bending moment distributions, it is usually necessary to consider a number of cases in order to determine the most critical. However, unless there is a change of cross-section within the length of the uprights, it is generally sufficient to concentrate on the lowest storey.

9.8 Design of the beams

The design of the beams in a pallet rack is little different than the design of any other cold-formed section beam. The beam ends do, of course, have semi-rigid joints but the consequences of this, in terms of both bending moment distribution and deflection, are taken into account in the down-aisle global analysis. The maximum design bending moment is generally near mid-span and primary beam design involves equating this maximum moment to the design moment of resistance of the beam. It may be noted here that, bearing in mind the non-linear moment-rotation behaviour of the beam end connectors, the FEM code allows up to 15% redistribution of end bending moments, as shown in Fig. 20.

Lateral torsional buckling is resisted to some extent by the pallets which provide the load. If, as is generally the case, the beams are of box section with (approximately) two axes of symmetry as shown in Fig. 7(a), the moment of resistance can be readily calculated and it is not necessary to test. The moments of resistance of the beams shown in Figs. 7(b) and (c) are best determined by a simple bending test which replicates the loading conditions expected in practice.

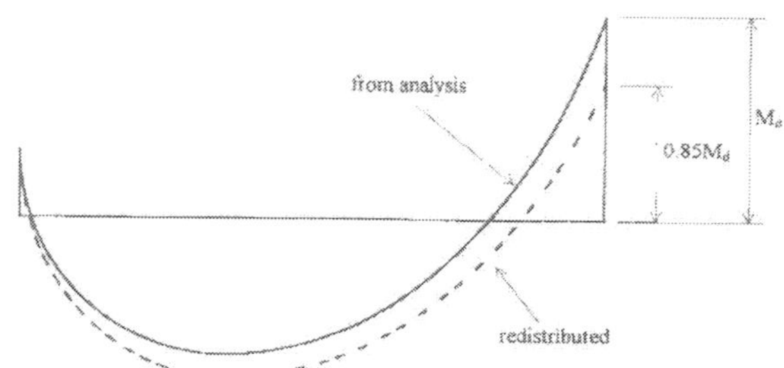

Figure 20. Redistribution of beam bending moments

9.9 Design of the beam to upright connectors

As the design bending moments have been determined in the global analyses and the design strengths have been determined by tests, the final design check of the beam to upright connectors is trivial. It should be noted that they are subject to the same redistribution of bending moments as the beams that they connect.

9.10 Design of the bracing system in the upright frames

Having determined the forces in the members (and connections) of the bracing system in the upright frames in the global analyses, their design is conventional. The bracing members are usually all of a single slender channel section and it is the compressive forces that govern the design.

References

AISI (1999) "Specification for the design of cold-formed steel structural members", *American Iron and Steel Institute*, 1996, including Supplement 1, Sept. 1999.

Davies J M (1992) "Down-aisle stability of rack structures", *11th International Speciality Conf. on Cold-Formed Steel Structs.*, St Louis, USA, October, 417-435.

Davies J M and Godley M H R (1998) "A European Design code for pallet racking",*14th International Speciality Conf. on Cold-Formed Steel Structs.*, St Louis, USA, October, 289-310.

EC3 (1996) Eurocode 3: Design of Steel Structures: Part 1.3: General Rules: Supplementary rules for cold formed thin gauge members and sheeting. *European Committee for Standardisation*, ENV 1993-1-3,

FEM (1999) "The design of steel static pallet racking and shelving*", Federation Europeenne de la Manutention, Document 10,2,02.*

Godley M H R (1991) "Storage racking", Chapter 11 of *Design of Cold Formed Steel Members*, Ed J Rhodes, Elsevier Applied Science, 361-399.

Hancock G J (1998) "Design of cold-formed steel structures (to Australian/New Zealand Standard AS/NZ 4600: 1996*), Australian Institute of Steel Construction*, CAN 00 973 839, 3rd Edition.

Hancock G J, Murray T W, Ellifritt D S (2001) "Cold-formed steel structures to the AISI Specification*", Marcel Dekker Inc.*, New York.

Horne M R (1975) "An approximate method of calculating the elastic critical loads of multi-storey plane frames*", Structural Engineer*, Vol. 53, No. 6, June.

RMI (1997) "Specification for the design, testing and utilization of steel storage racks", *Rack Manufacturers Institute, Charlotte NC.*